Springer Theses

Recognizing Outstanding Ph.D. Research

Aims and Scope

The series "Springer Theses" brings together a selection of the very best Ph.D. theses from around the world and across the physical sciences. Nominated and endorsed by two recognized specialists, each published volume has been selected for its scientific excellence and the high impact of its contents for the pertinent field of research. For greater accessibility to non-specialists, the published versions include an extended introduction, as well as a foreword by the student's supervisor explaining the special relevance of the work for the field. As a whole, the series will provide a valuable resource both for newcomers to the research fields described, and for other scientists seeking detailed background information on special questions. Finally, it provides an accredited documentation of the valuable contributions made by today's younger generation of scientists.

Theses may be nominated for publication in this series by heads of department at internationally leading universities or institutes and should fulfill all of the following criteria

- They must be written in good English.
- The topic should fall within the confines of Chemistry, Physics, Earth Sciences, Engineering and related interdisciplinary fields such as Materials, Nanoscience, Chemical Engineering, Complex Systems and Biophysics.
- The work reported in the thesis must represent a significant scientific advance.
- If the thesis includes previously published material, permission to reproduce this must be gained from the respective copyright holder (a maximum 30% of the thesis should be a verbatim reproduction from the author's previous publications).
- They must have been examined and passed during the 12 months prior to nomination.
- Each thesis should include a foreword by the supervisor outlining the significance of its content.
- The theses should have a clearly defined structure including an introduction accessible to new PhD students and scientists not expert in the relevant field.

Indexed by zbMATH.

More information about this series at https://link.springer.com/bookseries/8790

Johannes Sachs

Motion, Symmetry & Spectroscopy of Chiral Nanostructures

Doctoral Thesis accepted by
the University of Stuttgart, Germany

Springer

Author
Johannes Sachs
Max Planck Institute for Intelligent Systems
Stuttgart, Baden-Württemberg, Germany

Supervisor
Prof. Peer Fischer
Institute of Physical Chemistry
University of Stuttgart
Stuttgart, Germany

ISSN 2190-5053 ISSN 2190-5061 (electronic)
Springer Theses
ISBN 978-3-030-88688-2 ISBN 978-3-030-88689-9 (eBook)
https://doi.org/10.1007/978-3-030-88689-9

This Springer imprint is published by the registered company Springer Nature Switzerland AG
The registered company address is: Gewerbestrasse 11, 6330 Cham, Switzerland

Supervisor's Foreword

This Ph.D. thesis by Johannes Sachs discusses the dynamics and locomotion of colloidal structures in the context of their symmetry, and it shows how the Brownian motion of handed (chiral) nanostructures in solution enables a new spectroscopic observable.

Micro- and nanocolloids describe an important field in chemistry and physics. They are scientifically very interesting, e.g. as mesoscopic model systems with which atomic processes can be studied and visualized, and in their versatile applications. However, the vast majority of all colloids are spherical and are only used as passive "Brownian" particles. In contrast, micro- and nanocolloids that can be moved actively and in a controlled manner promise new applications, including the targeted transport of pharmaceuticals.

A special form of actively moving micro- and nanostructures are colloids in the shape of a cork-screw that possess a permanent dipole moment. In recent years, it has been possible to move these colloids through biological fluids, including through mucus and the vitreous of the eye. Because these microscrews have a magnetic moment that is perpendicular to the major axis, a rotating magnetic field can wirelessly rotate the screw such that it moves forward. Interestingly, such structures resemble the flagella of a bacterium in their locomotion. However, since the fabrication of chiral structures is rather complex, the question arises whether a screw really has to be handed (chiral), or whether it is possible to observe locomotion via rotation-translation coupling in a simpler structure that has higher symmetry. This thesis presents a rigorous symmetry analysis and it describes which dipolar structures can be propelled with a magnetic or an electric field. It also includes the demonstration of the first propeller that is not chiral, but that can still propel forward when it is rotated. The structure is an achiral "V"-shape and most likely the simplest propeller to date.

In Part II of his dissertation, Dr. Johannes Sachs has succeeded in observing a single chiral nanostructure in solution in such a way that he can determine the handedness of the nanostructure, i.e. if it is left-handed or right-handed. Probing single nanostructures promises applications with enhanced sensitivity and selectivity.

Generally, all optical methods used to determine the handedness of a sample require large numbers of molecules or colloids in solution. A difference of the solution's response to left- and right-circularly polarized light is a true measure of handedness/chirality. Because the large number of molecules or colloids are randomly oriented in solution, all polarization artefacts naturally vanish and interpretation of the spectral data becomes easy. However, this is no longer the case, if only one nanostructure is probed. Depending on its orientation the difference between left- and right-handed circularly polarized light is then not necessarily a measure of its chirality. Dr. Sachs has overcome this difficulty by observing a single nanostructure in a dark-field microscope and measuring its interaction with circularly polarized light over time as it reorients due to its Brownian motion. The time-averaged movement of the nanostructure then becomes a true measure of chirality. This is known as the ergodic principle, which has never before been applied to chiral spectroscopy. Dr. Sachs has thus succeeded in measuring a new observable: The rotationally averaged chiroptical spectrum of a single nanoparticle. Accessing this observable is an important advance that can pave the way for future studies of the (enhanced) interaction of nanostructures with chiral molecules.

The significance of the thesis does not only lie in the several firsts reported by Dr. Sachs but also that it beautifully connects fundamental questions to important applications. The motion of nanostructures cannot be understood without considering the underlying symmetries. Chiroptical signals cannot be interpreted without an understanding of the nanostructure's symmetry and that of the measurement setup. In his Ph.D. thesis, Dr. Sachs has provided novel insights and used these to realize original propulsive micro- and nanostructures as well as a new spectroscopic tool. It is thus well deserved, that this thesis is recognized with a Springer dissertation award.

Stuttgart, Germany
July 2021

Peer Fischer

Abstract

Nanostructures are of interest for a broad spectrum of potential applications. For biomedical tasks, small nanoswimmers could help realize targeted drug delivery or minimally invasive surgery, which necessitates a comprehensive understanding and control of their motion. As an optical sensor, single plasmonic nanostructures can enhance the weak optical signals associated with biologically important handed (chiral) molecules, and thus potentially lead to much higher detection sensitivities and improved selectivities. Both, the motion and spectroscopic behavior of nanostructures, are closely related to the symmetry they possess. For propulsion at small scales, it is well known that symmetry is important. For instance, the bacterial flagellum has a chiral corkscrew shape to allow non-reciprocal motion. One can therefore wonder if chirality is essential, and if indeed this is the simplest shape for propulsion, or if other structures—possibly ones that are simpler to fabricate—can also show propulsion when they are rotated. Another aspect that relates to the symmetry is the spectroscopic observation of nanostructures. Especially in chiral structures both the shape itself as well as the orientation in space determines optical signals. It is therefore important to be able to dis-entangle these effects.

To shed light on this, this thesis presents experiments that capture the active motion and chiroptical spectroscopy of artificial nanostructures at low Re, and examines the role chirality plays in these phenomena. Against previous expectations and reports, it is shown that a propeller does not need to be chiral to locomote and the first truly achiral propeller at low Re is reported. Similarly, a careful examination of the spectroscopy of plasmonic nanostructures shows that structures do not need to be chiral to give rise to chiroptical signals. A novel spectrometer was constructed to observe individual nanostructures. Thereby, it has been possible to identify an observable, which has not been measured previously—that permits the true chiroptical spectrum to be obtained from a single nanoparticle suspended in solution. The basis for the presented experiments is the fabrication of micro- and nanostructures with simple and complex, achiral and chiral shapes via GLAD.

Part I of this thesis considers active motion exhibited by highly symmetric micro- and nanoswimmers. Owing to the scallop theorem, reciprocal motion does not lead to a net translation at low Re, and other swimming strategies must be exploited, e.g. rotation-translation coupling. Previously, chirality was assumed to be required to efficiently couple rotation to translation at low Re, as is demonstrated by the corkscrew-shaped bacterial flagellum. However, a symmetry analysis suggested that much simpler shapes are potentially also propulsive if they are rotated in a non-trivial way. This would have important implications as then a novel class of micro- and nanoswimmers with highly symmetric, and easier to fabricate shapes, become feasible. Here, the propulsion characteristics of V-shaped objects that are driven by means of an external torque are investigated. The torque was either exerted by an external magnetic or an electric field and the "V"-shaped particle had a corresponding dipole moment. Experiments with macroscopic as well as microscopic rigid bodies revealed that the orientation of the dipole with respect to the body plays a crucial role for their ability to convert a rotation into a translation. Symmetry arguments are developed, which accurately predict whether or not the object is propulsive at all, and additionally if it moves uni- or bidirectionally. It is thereby unequivocally shown by theory and experimental evidence that chirality is not a prerequisite for efficient rotation-translation coupling at low Re but that propulsive objects are necessarily chiral if they are driven magnetically. Surprisingly, rotation of a truly achiral object will also lead to a propulsion, which is experimentally demonstrated for the first time utilizing an electrically driven "V"-shaped particle. Because chirality, i.e. $\widehat{P}$-odd symmetry, is not a requirement, an extended analysis that also includes the object's symmetry under charge conjugation ($\widehat{C}$) is employed to explain and predict the correct propulsion characteristics of arbitrary shaped objects by means of rotation-translation coupling.

In contrast, spherical objects are unable to utilize rotation-translation coupling. However, beads with unequal chemical reactivities on the two "faces" of their spherical body, so-called Janus particles, can show enhanced diffusion if a suitable chemical substrate is provided. A catalytic reaction converts the intrinsic asymmetry of the particles into a pressure gradient, which leads to a self-generated motion. Most reports of self-propulsion to date concern larger microparticles. In this thesis, it is examined if this means of propulsion can also be realized in sub-micron-sized Janus particles. The exact propulsion characteristics are not fully clear in most chemically driven particles and how their enhanced diffusion scales with size. Because of their small size, the observation of individual particles is not possible and thus high-density particle ensembles are observed by light-scattering techniques, i.e. DLS, DDM and SH-LDV. In contrast to DLS and DDM, SH-LDV is identified as a versatile and suitable technique for the characterization of active motion exhibited by Janus nanoparticles. First results on their size-dependent enhanced diffusion are presented, which show that this form of symmetry-breaking is effective at the smallest of scales.

Part II of this thesis considers the spectroscopic response exhibited by chiral- and achiral-shaped nanoparticles measured in a novel single-particle spectrometer. It is based on dark-field microscopy and contains a balanced detection setup, which was built to record the CDSI of a single nanoparticle, i.e. the difference in scattering intensities for left- and right-circularly polarized light. This is known as chiroptical spectroscopy and typical setups are limited to the examination of stationary single

Acknowledgements

I would like to thank all the people that contributed and helped me during the course of this thesis. My special appreciation goes to…

- my supervisor *Peer Fischer* for giving me the chance to work in his lab on interesting topics, and especially for giving me the opportunity to work out this thesis. Particularly, I want to thank him for opening up fantastic research collaborations and the possibility to attend inspiring conferences. He was undoubtedly the best imaginable adviser that has always been available for help and discussions. Furthermore, he is certainly an amazing scientist and a great personality.
- the two other examiners Prof. Dr. rer. nat. *Michael Börsch* and Prof. Dr. rer. nat., Dr. h. c. *Guido Schmitz* for agreeing to sit on the examination committee and take the effort to read and assess this thesis.

Further, I want to emphasize the research collaborators that took part in the theoretical and experimental work. They not only helped me in the lab but also with fruitful discussions which paved the way for this thesis and thus I want to especially thank…

- Prof. *Alexander M. Leshansky* and his colleagues *Konstantin I. Morozov* and *Oded Kenneth* from the Technion in Haifa for their mentoring regarding the theory of the propellers.
- Prof. *Thomas Palberg* and *Denis Botin* for their tremendous support throughout the SH-LDV experiments that took place in Mainz.
- Prof. *Thomas Sottmann* and *Kristina Schneider* for kindly granting access and helping with the DLS setup at University of Stuttgart.
- *Nico Segreto, Tian Qiu, S. Nikhilesh Kottapalli, Jan-Philipp Guenther, Cornelia Miksch, Andrew G. Mark, Jutta Hess* and all the other members from the MPI-IS Stuttgart for all the help they provided in and around our lab.

Finally, I need to express my deepest gratitude to my beloved wife *Bianca* and my son *Toni*. Thank you for supporting me and being there all the time!

Contents

Abbreviations

CB	Circular Birefringence
CCD	Charge-Coupled Device
CD	Circular Dichroism
CDSI	Circular Differential Scattering Intensity
CP	Circularly Polarized
DDM	Differential DynamicMicroscopy
DLS	Dynamic Light Scattering
FFT	Fast Fourier Transform
GLAD	Glancing Angle Deposition
JP	Janus Particle
LB	Linear Birefringence
LCP	Left-Circularly Polarized
LD	Linear Dichroism
LH	Left-Handed
LP	Linearly Polarized
LSPR	Localized Surface Plasmon Resonance
MSD	Mean Squared Displacement
NA	Numerical Aperture
OA	Optical Activity
OR	Optical Rotation
PVD	Physical Vapor Deposition
QWP	Quarter Waveplate
RCP	Right-Circularly Polarized
Re	Reynolds number
RH	Right-Handed
SEM	Scanning Electron Microscopy
SH-LDV	Super-Heterodyne Laser Doppler Velocimetry
TTL	Transistor-Transistor-Logic

Chapter 1
Introduction

Chirality is the property that defines an object that lacks mirror-symmetry. It is a fundamental property closely associated with life on earth and it therefore has far-reaching consequences. The fact that the world around us discriminates between left- and right-handed enantiomers is well-recognized but was certainly not envisioned by Louis Pasteur when he started the first experiments with chiral molecules and crystals more than 170 years ago. However, today we know that almost all biomolecules are chiral. Remarkably, sugars, amino acids, nucleic acids and many other biomolecules occur almost exclusively with one handedness in nature. The question why homochirality exists, touches on the question of the origin of life. For example, only one of the two mirror-image forms of sugars and/or nineteen of the twenty amino-acids can be enzymatically digested by the human body and therefore chirality is highly important for biochemistry and pharmaceuticals. In practice, the handedness of a substance is determined by the unique optical signature chiral molecules have in solution.

The structures of molecules, proteins, and other building blocks of life typically have a size about a nanometer and smaller. They experience passive (Brownian) diffusion in solution, which is, however, not very efficient for transport over larger distances. A small protein that diffuses in water will still need tens of years to cover a root-mean-square distance of a meter. Thus, biology has devised more effective transport mechanisms to generate active motion at the smallest of length scales. Consequently, naturally occurring nanomotors or nanoswimmers that enable directed transport are of utmost importance. However, biological machines also play a role on larger length scales, such as the corkscrew-shaped bacterial flagellum that allows propulsion at small Reynolds numbers. Since Richard Feynman gave his seminal lecture "There is plenty of room at the bottom" in December 1959 [1] the quest to build man-made micro- and nanomachinery that may one day rival biological nanomotors has attracted considerable attention.

J. Sachs, *Motion, Symmetry & Spectroscopy of Chiral Nanostructures*, Springer Theses,
https://doi.org/10.1007/978-3-030-88689-9_1

In this thesis, artificial micro- and nanostructures are considered and it investigates (1) how fundamental symmetries relate to their motion and (2) how a chiral shape impacts their optical detection when they are motile.

While the lack of symmetry is exploited for optical detection, the presence of symmetry has important consequences for Brownian motion and propulsion of small objects if they reside in fluidic environments at low Re, which is the flow regime relevant for micro- and nanoswimmers. It is widely known that, owing to fundamental physical principles, symmetry has to be broken to enable efficient propulsion at low Re (in the case of force-free swimming)—a reciprocal motion does not lead to a net translation in such situations. Therefore, bacteria utilize chiral-shaped (symmetry-broken) flagella in order to propel. By spinning their corkscrew-like flagellum they convert rotation into a translation of their body. This work examines if geometrical chirality is a pre-requisite for such corkscrew-like motion, or if much simpler, more symmetrical and achiral shapes, can also be propelled by rotation-translation coupling. It thereby explores how a dipole moment impacts the overall chirality of an object and if a propeller in general has to be chiral in order to efficiently swim in the laminar flow regime. To address this question experimentally, highly symmetrical micro- and nanostructures were fabricated and thereby the propulsion characteristic in dependence on the objects' fundamental symmetries was examined. Remarkably, it was found that a truly achiral object can also propel by rotation-translation coupling, and this led to the realization of the first achiral propeller of its kind that propels in the laminar flow regime.

Another class of even more symmetrical structures is examined next: spherical particles that propel by means of self-generated concentration gradients. A spherical object shows no rotation-translation coupling owing to its isotropic body. However, artificial Janus microparticles, i.e. spherical beads consisting of two hemispheres with different physio-chemical properties, can (in the presence of a suitable fuel) generate a local chemical gradient that in turn causes active motion. Although many proof-of-principle experiments with such microscale particles exists, the exact nature of the underlying propulsion mechanism is in many examples still not fully determined. In particular, there is still uncertainty if the enhanced diffusion clearly observed at micron-scales can also cause enhanced diffusion of nanometer-sized particles. To address this question, the enhanced diffusion of symmetry-broken spherical Janus *nano*-particle ensembles is examined. Due to their small sizes compared to the wavelengths of visible light, their observation cannot be realized by traditional microscopy techniques. Instead, spectroscopic methods are necessary to measure their diffusion. This thesis describes various light scattering techniques that were employed to study the enhanced diffusion in ensembles of sub-micron chemically-powered swimmers.

While scattering and spectroscopy of ensembles is fairly well established, the spectroscopic observation of single nanoparticles that are suspended in solution remains challenging. It would enable interesting perspectives for future applications. Especially plasmonic resonant nanostructures, if they can be spectroscopically observed, could facilitate the optical detection of molecular chirality. The latter typically yields weak signals owing to the small sizes of molecules, whereas nanoparticles exhibit a significantly increased interaction strength with visible light owing to their size and plasmonic activity. Chiroptical spectroscopy detects the unequal interaction of a sample with left- and right-circularly polarized light. In general, achiral samples should display no differential intensities in their chiroptical response. However, it is accepted that this is only true for an isotropic medium, i.e. for an ensemble in a cuvette, since even achiral, but oriented, samples can give rise to strong circular intensity differences. In this case, a circular intensity difference is not a measure of the sample's intrinsic chirality, but rather a consequence of externally induced symmetry-breaking.

Therefore, the second part of this thesis examines the impact of the light-object symmetry on a single nanoparticle's chiroptical response. Single-particle (chiroptical) spectroscopy is the subject of intense research efforts, but to date only immobilized stationary molecules and nano-objects have been measured. Almost always, the particles are fixed to a surface, which is particularly problematic, as the particle-surface interaction already introduces an asymmetry that can mask any true chiral signals. Here, a novel detection setup is used to investigate, for the first time, a single nanoparticle that is freely suspended in solution, far away from a surface, and thereby randomly reorients due to Brownian motion. If such a particle is observed for long enough times, all possible orientations in space are sampled isotropically and consequently a true chiroptical signal can be detected, revealing the particles' intrinsic chirality. This approach is shown to yield a spectral response which is equivalent to traditional bulk ensemble spectroscopy, but requires only one single nanoparticle. The equivalence between a single-particle and an ensemble measurement is in accord with the ergodic hypothesis and consequently the same information is provided by both measurements. However, observing single nanostructures permits measurements with very low sample volumes, compared to traditional chiroptical spectroscopic techniques.

The applications of the micro- and nanostructures considered in this thesis range from biomedical tasks related to transport to novel sensing applications. The profound understanding of their propulsion and spectroscopic response is a key prerequisite for their practical use. Symmetries can be utilized to support future design principles. Also, a novel observable that permits the detection of the intrinsic chirality for single plasmonic nanoparticles has been established. It thereby brings the field closer to the detection of chiroptical spectra from a single molecule. Altogether, the motion, symmetry and spectroscopy of chiral nanostructures was and is subject to highly interdisciplinary research efforts across all areas of science, from physics and material science, over chemistry and medicine to biology. This thesis has made fundamental contributions to the motion as well the observation of chiral micro- and nanostructures in solution.

Reference

1. Feynman RP (1960) There's plenty of room at the bottom. California Institute of Technology, Engineering and Science Magazine

Chapter 2
Fundamentals of Chiral Nanostructures in Fluids

In this chapter the theoretical background concerning the behaviour of small scale objects in external electromagnetic fields and in fluid environments are introduced. For the subsequent work of the thesis results from classical hydrodynamics (Sects. 2.1–2.3) to mathematical concepts (Sect. 2.5) and a description how the nanostructures are prepared (Sect. 2.7), are needed. The topics are only discussed briefly as further details can be found in many textbooks, including [1–3] where the more interested reader can find further information.

2.1 The Navier–Stokes Equation

The flow of fluids and motion in fluidic environments is derived from the fundamental conservation laws of physics. As the fluid density must be conserved over time the local density at a distinct point or stationary volume element can only change if it is associated with a flow of mass. Mathematically this is described by the continuity equation [1]

$$\frac{\partial}{\partial t}\rho + \nabla \cdot (\rho \mathbf{u}) = 0, \tag{2.1}$$

where ρ is the local density and $\mathbf{u}$ is the average velocity of the associated mass flow. Moreover it was assumed that the energy in the system is conserved, and that there are no sources or sinks, as otherwise the right hand side of Eq. (2.1) would be $\neq 0$. By further assuming an incompressible fluid with constant density ρ, which is a valid assumption for water and since sound propagation will not be considered, the equation of continuity further reduces to

$$\nabla \cdot \mathbf{u} = 0, \tag{2.2}$$

J. Sachs, *Motion, Symmetry & Spectroscopy of Chiral Nanostructures*, Springer Theses,
https://doi.org/10.1007/978-3-030-88689-9_2

stating that a fluid's mass flow rate through an area is constant. In other words, if water is flowing through a channel and the cross-section of this channel is varied, the fluid flow velocity must also change. For example if a pipe's diameter is reduced (increased), this leads to a faster (slower) motion of the fluid to make sure the same amount of water is flowing per time. This acceleration or slowing down corresponds to external forces acting on the fluid, according to Newton's second law of motion $\mathbf{F} = \frac{d\mathbf{p}}{dt} = m \cdot \mathbf{a}$, where $m = \rho \Delta V$ is the mass of a volume element ΔV with density ρ and $\mathbf{a}$ is the acceleration.

The application of Newton's law of conservation of momentum to a volume of fluid leads to:

$$\mathbf{F}_{\textbf{total}} = \rho \Delta V \frac{D}{Dt} \mathbf{u}(\mathbf{x}, \mathbf{t}) = \mathbf{F}_{\textbf{pressure}} + \mathbf{F}_{\textbf{friction}} + \mathbf{F}_{\textbf{external}}. \tag{2.3}$$

This force balance simply states that the rate with which momentum of a small volume element changes must equal the surface and external forces (e.g. gravity). The corresponding expressions for the terms on the right hand side together with the material differential $\frac{D}{Dt}$ on the left hand side, Eq. (2.3) give:

$$\rho \left(\frac{\partial}{\partial t} \mathbf{u} + \mathbf{u} \cdot \nabla \mathbf{u} \right) = -\nabla p + \eta \nabla^2 \mathbf{u} + \rho \mathbf{a}, \tag{2.4}$$

which is the Navier–Stokes-equation [1] for incompressible fluids with constant (dynamic) viscosity η, pressure p and acceleration $\mathbf{a}$ due to external forces. The last term summarizes external forces and is here explicitly written for gravity ($\mathbf{a} = \mathbf{g}$). Although it is one of the most general equations in the field of fluid dynamics it is in most cases not possible to find analytical solutions. Therefore it is helpful if the equation can be simplified.

An approximation to the Navier–Stokes-equation can be made in cases where the inertial terms are dominant. By neglecting the viscous terms in Eq. (2.4) one can derive the so called Euler equation, which will not be discussed here, as it is not relevant for the work presented herein.

For small scale objects and nanostructures a simplification can be made that is relevant for this thesis. Because of their small size their volume and mass is negligible, thus inertial forces on the particle play no role and the associated terms in Eq. (2.4) can be omitted ($\rho \left(\frac{\partial}{\partial t} \mathbf{u} + \mathbf{u} \cdot \nabla \mathbf{u} \right) \sim 0$ and $\rho \mathbf{a} \sim 0$). Together with Eq. (2.2) this leads to the so called creeping motion or Stokes equations [1]:

$$\begin{aligned} -\nabla p + \eta \nabla^2 \mathbf{u} &= 0, \\ \nabla \cdot \mathbf{u} &= 0, \end{aligned} \tag{2.5}$$

which describes the motion of a small object in a fluid. There are two important consequences of the simplification made in the Stokes equations. First, it is a set of differential equations that are linear in velocity $\mathbf{u}$ and pressure p. This opens up a way to analytically solve the equations by using classical superposition techniques, even in the case of complex geometries. Second, and even more important for the subject of this thesis, is that the time derivative was dropped too. Therefore, there is no longer any explicit time dependence in a set of solutions ($\mathbf{u}_1$, p_1) implying that ($-\mathbf{u}_1$, $-p_1$) also satisfies Eq. (2.5). Thus, the Stokes equations and any solution is time-reversible. This finding drastically impacts the motion of a nanostructure in a fluid and put fundamental constraints on the design of artificial microswimmers, as discussed in the next section.

2.2 Motion in Low Reynolds Number Environments

The Stokes equations 2.5 describe the motion of an object as long as the approximations made to the Navier–Stokes-equation are valid. This is the case if the influence of inertial terms ($\rho\mathbf{u}\cdot\nabla\mathbf{u}$) are much smaller than the viscous terms ($\eta\nabla^2\mathbf{u}$). The ratio of the two is known as the Reynolds number (Re):

$$\text{Re} = \frac{\boldsymbol{F}_{\text{inertial}}}{\boldsymbol{F}_{\text{viscous}}} \propto \frac{\rho\mathbf{u}\cdot\nabla\mathbf{u}}{\eta\nabla^2\mathbf{u}}, \tag{2.6}$$

which is a dimensionless quantity. Thus, for low Reynolds numbers (Re < 1) environments the Stokes equations provide a good approximate solution to the Navier–Stokes-equation. By identifying L as a characteristic length scale of an object ($\nabla \propto \text{L}^{-1}$) and U as a characteristic translational velocity of the fluid ($\mathbf{u} \propto \text{U}$) an alternative definition would be:

$$\text{Re} \propto \frac{\rho\text{U}^2\,/\,\text{L}}{\eta\text{U}\,/\,\text{L}^2} = \frac{\rho\text{L}\,\text{U}}{\eta}. \tag{2.7}$$

in which ρ is the density of the object and η is the dynamic viscosity of the medium. Now one can rapidly approximate whether or not low Re conditions are given for a certain problem. It is easy to verify that nanostructures with a characteristic length of $\text{L} \sim 10^{-9}$ m operate at low Re conditions if they are immersed in an aqueous solution with $\rho\,/\,\eta \sim 10^6\ \text{sm}^{-2}$. Because V typically is on the order of a body length per second, the experiments that are reported in this thesis fall under the regime of low Re numbers and their theoretical description will therefore refer to the solution of Eq. (2.5). If a rotational motion is assumed one has to consider that the rotational speed of the fluid is $\omega = U\,/\,L$. Thus, the rotational Reynolds number would be $\text{Re}^{\text{rot}} = \rho\text{L}^2\,\omega\,/\,\eta$.

As mentioned, omitting the inertial terms at low Re leads to time-reversibility and puts substantial constraints on the possible mechanisms that can be used for swimming of a small scale object. A vivid description of what the time-reversible symmetry means was given by Nobel laureate Edward Purcell [4] with his famous scallop-theorem in which he explained a scallop's motion at low Re. When the scallop opens its arms it will move forward, but in the moment it closes them again it moves backwards. In fact, this imagined scallop traverses along the exact same trajectory in both cycles, just with a reversed direction, thus it will end up in the same position where it started. Because there is no time-dependence in Eq. (2.5) time plays no role in the motion and it cannot make a difference if the opening or closing is fast or slow. He demonstrated that the opening and closing cycle is a reciprocal motion and the net translation of such a reciprocal motion in Newtonian fluids, that is a fluid with constant viscosity, will always be zero.

For non-Newtonian fluids by contrast, the time-dependence in the Navier–Stokes-equation can not be omitted anymore and it does make a difference how fast or slow the scallop opens or closes its arms. Recently it has been demonstrated that swimming by reciprocal motion is possible if the shear rate of the viscous fluid changes with time [5]. This was done by actuating an artificial scallop with an external magnetic field, which made it possible to control the speed of the opening and closing cycles.

However, in Newtonian fluids one needs to break time-reversible symmetry to achieve net translation. To do that, a non-reciprocal motion could be used in order to generate propulsion. One approach is to continuously rotate an object with the same rotation sense and exploit the torque-velocity coupling which is inherent to the object's shape. This mechanism is known as the propeller effect [6] and the classical model for this mechanism is corkscrew motion, where the corkscrew moves forward by turning around its long axis. Note that the handedness of the corkscrew fixes the propulsion direction, i.e. a right-handed corkscrew is going forward when turned clockwise while a left-handed screw would go the other way. This behaviour is commonly known as unidirectional propulsion and an intriguing example of this propulsion mechanism can be found in nature when bacteria spin their flagella in a corkscrew-like stroke to generate locomotion. Artificial corkscrew-like propellers mimicking this behaviour have been demonstrated as well [7–9]. For this, rigid magnetic nano- or microscrews were fabricated and were rotated by an external magnetic field. Again, by exploiting the shape inherent rotation-translation-coupling they were able to move through water and even biological tissue [10, 11]. These are strictly not swimmers as an external torque is applied, and so this approach will be denoted as *physically powered* propulsion mechanism throughout this thesis. While the corkscrew has a low symmetry and intuitively couples rotation to translation the propulsion mechanism of highly symmetrical rigid nanostructures is investigated in Chap. 3.

There are also other strategies to generate non-reciprocal motion patterns. In nature microorganisms make use of flexible structures. For instance, they move thousands of cilia on their surface in a spatially and temporally coordinated manner. This creates a metachronal (i.e. travelling) wave around their body, which leads to locomotion. A simplified variant of this complex behaviour was exploited for an artificial swimmer that moves by deforming its shape [12]. The object consists of a flexible elastomer that expands upon irradiation with light. By scanning light over the body with a periodic, but non-reciprocal pattern the body is deforming its shape and can travel with a wave-like gait. However, this thesis is only dealing with rigid body swimmers and the focus herein is on the question whether structures other than the complex and low-symmetry corkscrew are able to propel at low Re and if they can be controllably steered.

A sphere is a highly symmetrical shape and can therefore never be propulsive by means of (physically powered) rotation-translation-coupling in a Newtonian fluid (see Sect. 2.3). In order to generate locomotion the symmetry of the problem has to be broken by another mechanism. The idea to overcome the limitations of the time-reversibility in Eq. (2.5) in this case is to use chemical gradients. Thus, this approach is termed *chemically powered* propulsion, an approach inspired by molecular motors. The artificial model is usually realized by means of a Janus particle (JP), which is an object that has two distinct physical or chemical properties across its surface. The name stems from the two-faced Roman god *Janus*. The most general form of a Janus particle (JP) is a sphere consisting of two halves made of different materials. This leads to different chemical reactivities on both ends. In presence of a substrate (or *fuel*) one half is catalytically active while the other is not, creating chemical reaction products on one side only. Sometimes other mechanisms are used to create the gradient, e.g. a temperature difference by exploiting a mismatch in the material's ability to absorb light. Regardless of the mechanism, an osmotic pressure difference is generated across the surface of the sphere. According to the Stokes-Equations (Eq. (2.5)) a pressure gradient ∇p must drive a flow and hence, the fluid streams from one half of the sphere to the other. In an incompressible fluid this flow moves the sphere forward in the opposite direction to the streaming. Because the motion is generated by the particle itself, this effect is referred to as self-propulsion. Experiments that investigate the self-propulsion of chemically powered JP are described in Sect. 3.6.

2.3 Rotation-Translation-Coupling at Low Re

At low Re the principle of coupling rotation to translation for an arbitrarily shaped object is described by the linear differential Eqs. 2.5. The solution is therefore also linear and in the moment a particle is rotating or translating the fluid will exert a frictional drag force $\boldsymbol{F}_d$ and torque $\boldsymbol{L}_d$ on it [13]:

$$\begin{pmatrix} \boldsymbol{F}_d \\ \boldsymbol{L}_d \end{pmatrix} = -\begin{pmatrix} \boldsymbol{A} & \tilde{\boldsymbol{C}} \\ \boldsymbol{C} & \boldsymbol{B} \end{pmatrix} \cdot \begin{pmatrix} \boldsymbol{U} \\ \boldsymbol{\Omega} \end{pmatrix} \tag{2.8}$$

Here $\boldsymbol{U}$ and $\boldsymbol{\Omega}$ are the translational and the rotation (angular) velocities. $\boldsymbol{A}$ is the translation resistance tensor, $\boldsymbol{B}$ the rotation resistance tensor and $\boldsymbol{C}$ the rotation-translation coupling tensor with the tilde denoting the transpose of tensors. If there is an external force $\boldsymbol{F}$ and torque $\boldsymbol{L}$ acting on the particle its induced motion is simply counterbalanced by the viscous forces and torques $\boldsymbol{F} = -\boldsymbol{F}_d$ and $\boldsymbol{L} = -\boldsymbol{L}_d$. The system of equations then reads as:

$$\begin{pmatrix} \boldsymbol{F} \\ \boldsymbol{L} \end{pmatrix} = \begin{pmatrix} \boldsymbol{A} & \tilde{\boldsymbol{C}} \\ \boldsymbol{C} & \boldsymbol{B} \end{pmatrix} \cdot \begin{pmatrix} \boldsymbol{U} \\ \boldsymbol{\Omega} \end{pmatrix}. \tag{2.9}$$

The square matrices in (2.8) and (2.9) are known as the resistance (or friction) matrix $\boldsymbol{T}$ as it contains the 3 resistance tensors: $\boldsymbol{A}$, $\boldsymbol{B}$, $\boldsymbol{C}$. The inverse is known as the mobility matrix $\boldsymbol{M}$ so that $\boldsymbol{T}\boldsymbol{M} = \mathbb{1}$, or:

$$\boldsymbol{M} = \begin{pmatrix} \mathscr{D} & \tilde{\mathscr{G}} \\ \mathscr{G} & \mathscr{F} \end{pmatrix} = \begin{pmatrix} \boldsymbol{A} & \tilde{\boldsymbol{C}} \\ \boldsymbol{C} & \boldsymbol{B} \end{pmatrix}^{-1} = \boldsymbol{T}^{-1}, \tag{2.10}$$

thus Eq. (2.9) becomes:

$$\begin{pmatrix} \boldsymbol{U} \\ \boldsymbol{\Omega} \end{pmatrix} = \begin{pmatrix} \mathscr{D} & \tilde{\mathscr{G}} \\ \mathscr{G} & \mathscr{F} \end{pmatrix} \cdot \begin{pmatrix} \boldsymbol{F} \\ \boldsymbol{L} \end{pmatrix}. \tag{2.11}$$

Consequently $\mathscr{D} = \boldsymbol{A}^{-1}$ is the translation mobility tensor, $\mathscr{F} = \boldsymbol{B}^{-1}$ the rotation mobility tensor and $\mathscr{G} = \boldsymbol{C}^{-1}$ is the coupling mobility tensor. All of these tensors are second-rank tensors and the translational and rotational tensors will always be symmetric ($\tilde{\boldsymbol{A}} = \boldsymbol{A}$, $\tilde{\boldsymbol{B}} = \boldsymbol{B}$). As Happel and Brenner have shown, there is a unique point for each object where the coupling resistance tensor $\boldsymbol{C}$ can be made symmetric too [1]. This is termed the hydrodynamic *center of resistance* and it can be shown that $\boldsymbol{C} = \tilde{\boldsymbol{C}}$ at this point. The same is true for the coupling mobility tensor, for which $\mathscr{G} = \tilde{\mathscr{G}}$ at the hydrodynamic *center of mobility*. It is important to note that those two points in general do not coincide [14].

The principal axes of a tensor form a basis of eigenvectors in which the tensor is diagonal, and the elements on the diagonal are the corresponding eigenvalues, called the principal values, e.g. the principal moments of inertia. The eigenvectors that diagonalize $\boldsymbol{A}$, $\boldsymbol{B}$ or $\boldsymbol{C}$ are called *principal axes of translation, rotation* and *coupling*, respectively. Conversely it is also possible to find the *principal axes* in the eigensystem of the mobility tensors $\mathscr{D}$, $\mathscr{F}$ and $\mathscr{G}$ [15]. Again, the *principal axes* in the two distinct reference frames are not necessarily equal. In Chap. 3 of this thesis the eigensystem based on the rotation mobility matrix is being used to describe the rotation-translation coupling of a V-shaped object.

If one imagines a particle with a dipole moment in an external electromagnetic field, then a gradient electromagnetic field would drag the dipole through the fluid along the field lines by exerting a force on it. For gradient-free fields, and under the assumption that there are no other external force (e.g. gravity), $\boldsymbol{F} = 0$. In this force-free case Eq. (2.11) reduces to:

$$\boldsymbol{U} = \mathscr{G} \cdot \boldsymbol{L}, \quad \boldsymbol{\Omega} = \mathscr{F} \cdot \boldsymbol{L}. \tag{2.12}$$

These are the governing equations for actuating a particle in a viscous fluid by means of an external torque. The translational velocity $\boldsymbol{U}$ is coupled to the torque $\boldsymbol{L}$ via the rotation-translation coupling mobility tensor $\mathscr{G}$. Its elements are connected to the symmetry of an object and is generally dependent on how the origin of coordinates is chosen. In the next Sect. 2.3.1 the form of $\mathscr{G} = \boldsymbol{C}^{-1}$ is shown for differently shaped objects.

2.3.1 Symmetry and the Coupling Tensor

The motion of an arbitrary object in a viscous fluid at low Re can be fully described by the 6×6 resistance matrix $\boldsymbol{T}$ of Eq. (2.10), which generally has 36 entries. The entries are subject to different tensors that have different transformation properties representing their physical nature. It can be proven that the translation and rotation resistance tensors $(\boldsymbol{A}, \boldsymbol{B})$ are symmetric and because of this symmetry they both carry only 6 independent elements [1]. The coupling tensor $\boldsymbol{C}$ has in general 9 entries and depends on the choice of origin, e.g. the resistance coupling tensor is symmetric at the hydrodynamic center of reaction. If a shape possesses additional (geometric) symmetries one can apply the corresponding symmetry operations on the resistance tensors and determine if any terms are equal or if they vanish. In doing so the number of the 21 remaining elements in the resistance matrix can often be reduced further. For this thesis it is particularly interesting how the entries of the rotation-translation coupling tensor transform.

In case of highly symmetrical shapes with three mutually perpendicular symmetry planes (orthotropic bodies), e.g. a sphere, cube or a polyhedron, the coupling tensor completely vanishes, $\boldsymbol{C} = 0$, and the two other resistance tensors reduce to $\boldsymbol{A} = A\mathbb{1}$ and $\boldsymbol{B} = B\mathbb{1}$. From Eq. (2.12) it readily becomes clear that a body with $\boldsymbol{C}^{-1} = \mathscr{G} = 0$ will not couple its rotational motion into a translational movement. Thus, such shapes will never be propulsive by means of rotation-translation-coupling. Or, viewed the other way round, there will never be a torque exerted on a particle when it is dragged through the fluid as long as $\boldsymbol{C} = 0$.

As mentioned above, an (infinite) corkscrew is a low-symmetry helicoidal object and one can prove that its resistance tensors are given by [1]:

$$\boldsymbol{A} = \begin{pmatrix} A_{11} & \cdot & \cdot \\ \cdot & A_{22} & \cdot \\ \cdot & \cdot & A_{33} \end{pmatrix}, \quad \boldsymbol{B} = \begin{pmatrix} B_{11} & \cdot & \cdot \\ \cdot & B_{22} & \cdot \\ \cdot & \cdot & B_{33} \end{pmatrix}, \quad \boldsymbol{C}_{\text{helix}} = \begin{pmatrix} C_{11} & \cdot & \cdot \\ \cdot & C_{22} & \cdot \\ \cdot & \cdot & C_{22} \end{pmatrix} \tag{2.13}$$

Due to the diagonal form of the tensors a rotation around one of the principal axes will lead to a translation, as is expected for a corkscrew propeller. Chapter 3 of this thesis tackles the problem of propulsion of rigid bodies at low Re that have a highly symmetrical geometry with two mutually perpendicular symmetry planes (reflection symmetries). Such an object can be realized by a V-shape (see Fig. 2.1) or arc-like object. One would probably expect that such an object can not really propel and certainly not in a controlled direction. One would probably also expect that such an object can not be chiral as it possess symmetry planes. It will be shown that there are several surprises, if one closely examines the underlying symmetries. In the case of a V-shape or an arc $\boldsymbol{A}$ and $\boldsymbol{B}$ are the same as in Eq. (2.13) and the coupling tensor is [1]:

$$\boldsymbol{C}_{\text{V-Shape}} = \begin{pmatrix} 0 & 0 & 0 \\ 0 & 0 & C_{23} \\ 0 & C_{32} & 0 \end{pmatrix} \tag{2.14}$$

Fig. 2.1 Visualization of a body that is shaped like a "V" and that posses two mutually orthogonal planes of mirror symmetry

Despite the high symmetry there are two non-zero entries remaining that could be exploited to couple such an object's rotation to its translation. Note, that here the resistance tensors ($\boldsymbol{A}$, $\boldsymbol{B}$, $\boldsymbol{C}$) are given at the hydrodynamic center of reaction. The form (vanishing and non-zero elements) of the mobility tensors ($\mathscr{D}, \mathscr{F}, \mathscr{G}$) will be the same if one chooses the hydrodynamic center of mobility as the coordinate origin, which is mainly done throughout this thesis.

2.4 Brownian Motion

In the complete absence of external forces and torques there can still be a motion observed for small scale particles like nanostructures. It is a random motion, stemming from thermal fluctuations also known as Brownian motion. Brownian motion occurs due to the finite kinetic energy all molecules and objects possess at thermal equilibrium. Random collision of a particle with solvent molecules creates a force on the particle and causes it to move. At a higher temperatures molecules have higher velocities and therefore exert higher forces and cause thus faster motion of the particle, representing the stochastic nature of this effect.

A particle's mobility in general depends on its shape, and is thus a tensorial quantity ($\mathscr{D}$ and $\mathscr{G}$), as described above in Sect. 2.3.1. It was already mentioned there, that for simple and highly symmetrical particles, like a sphere, the coupling tensor vanishes ($\boldsymbol{C} = 0$). Further, the translational and rotational friction tensors can be reduced to a single element because a sphere's mobility must be the same in any direction: $\boldsymbol{A} = A\mathbb{1}$, $\boldsymbol{B} = B\mathbb{1}$. Therefore, the problem is isotropic and can be described by single scalar quantities $|\boldsymbol{A}| = \mathrm{A}$, $|\boldsymbol{B}| = \mathrm{B}$. Consequently, the inverse mobilities are also scalars and conventionally they are denoted as $|\mathscr{D}| = \mu_T$ and $|\mathscr{F}| = \mu_R$, describing the translational and rotational (isotropic) mobility of a sphere. The connection between a particle's mobility and the random thermal collisions with molecules can now be deduced from the fluctuation-dissipation theorem and was

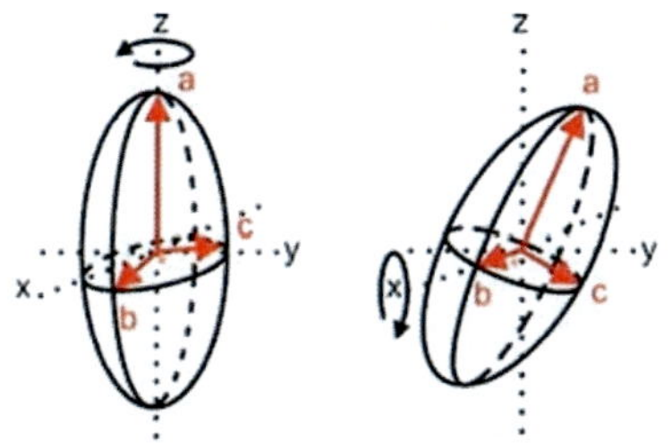

Fig. 2.2 Schematic visualization of the rotational diffusion a prolate ($a > b = c$) ellipsoid undertakes around the long (**a**) and short (**b**) semi-axes. Reprinted from Ref. [17], ©IOP Publishing. Reproduced with permission. All rights reserved

first explained by Einstein and Smoluchowski. The Stokes-Einstein (sometimes also Einstein-Sutherland) equation in its most general form reads:

$$D = \mu \, \mathrm{k}_B T, \tag{2.15}$$

and it relates a diffusion constant D to the hydrodynamic mobility μ and the temperature T, where k_B is the Boltzmann constant. Stokes derived the exact values of μ from Eq. (2.5) by calculating the flow around a sphere that is dragged through a fluid. The result is the well known Stokes' drag force $\boldsymbol{F}_{\text{drag}} = v\mu^{-1}$ with:

$$\mu_T = \frac{1}{6\pi\eta r}, \quad \mu_R = \frac{1}{8\pi\eta r^3}. \tag{2.16}$$

Here the subscript R denotes the rotational and T the translational mobility, r is the sphere's radius and η the viscosity of the surrounding solvent.

For more complex shapes the derivation gets more complicated because it is no longer an isotropic problem and thus it cannot be described by a single scalar anymore. For an elongated particle different translational and rotational mobilities along or around the long and short axes arise. For example, the diffusion of a prolate or oblate ellipsoid with semi-axes $|\mathbf{a}| = a$ and $|\mathbf{b}| = |\mathbf{c}| = b$ and volume $V = (4/3)\pi a \cdot b^2$ (see Fig. 2.2) was initially described by Perrin and later corrected by Koenig [16]. They derived correction factors $g_{\parallel}$ and $g_{\perp}$, which quantify the deviation of the ellipsoids rotational diffusion from the rotational diffusion a sphere with the same volume would exhibit. Thus, Eq. (2.15) will be different for the rotation around the long and short symmetry axes [17],

$$\mu_R^{\parallel} = \frac{1}{6\eta V g_{\parallel}}, \quad \mu_R^{\perp} = \frac{1}{6\eta V g_{\perp}}, \tag{2.17}$$

where $\mu_R^{\parallel}$ is the mobility around the long axis (Fig. 2.2a) and $\mu_R^{\perp}$ the mobility for a rotation around the short or equatorial axes (Fig. 2.2b) of the ellipsoid, and the correction factors are:

$$g_{\parallel} = \frac{2\left(p^2 - 1\right)}{3p(p - S)}, \quad g_{\perp} = \frac{2\left(p^4 - 1\right)}{3p\left[\left(2p^2 - 1\right)S - p\right]}. \tag{2.18}$$

Here, $p = a/b$ is the aspect ratio of the semi-axes and S is a factor that is different for prolate ($a > b$) and oblate ($a < b$) shapes:

$$\begin{aligned} S_{\text{prolate}} &= \frac{1}{\sqrt{p^2-1}} \ln[p + \sqrt{p^2 - 1}], \\ S_{\text{oblate}} &= \frac{1}{\sqrt{1-p^2}} \arctan\left[p^{-1}\left(1 - p^2\right)^{1/2}\right]. \end{aligned} \tag{2.19}$$

The translational diffusion is also different for a motion along or perpendicular to the long axis **a**, whereas the average is:

$$\mu_T = \frac{S_{\text{prolate}}}{6\pi\eta b}. \tag{2.20}$$

Those expressions might still be reasonable to work with but for other, even more complex and arbitrary shapes (e.g. molecules, bundles, aggregates, etc.) an exact solution is tedious. Therefore, the object is usually approximated as a sphere with the *hydrodynamic radius* r_{h} in order to preserve the compact expression from Eq. (2.16). This radius r_{h} is not an exact or realistic value, instead it resembles the radius of a spherical volume that would have the exact same hydrodynamic properties as the more complex object.

D is the diffusion constant and it was shown by Einstein that the root-mean-square distance $\langle\sqrt{\mathbf{x}^2}\rangle$ an object diffuses over time is linear and related to the diffusion constant by:

$$\langle\sqrt{\mathbf{x}^2}\rangle = \sqrt{2d \cdot D \cdot t}, \tag{2.21}$$

with t the observation time and d the number of dimensions (for 1D d=1; for 2D d=2, and for 3D d=3). The factor d takes into account that a confined motion will limit the randomness and irrespective of the confinement the net translation an object exhibits is on average zero. It follows that the diffusion constant D is:

$$D = \frac{\langle\mathbf{x}^2\rangle}{2d \cdot t}, \tag{2.22}$$

and the average diffusion observed for an ensemble of N particles is known as mean squared displacement (MSD):

$$\text{MSD} = \langle\mathbf{x}^2\rangle = \frac{1}{N}\sum_{i=1}^{N}\mathbf{x}_i^2 \tag{2.23}$$

The effect of Brownian Motion can be divided into two regimes. For long times the MSD is linear with t as expected fom Eqs. (2.22) and (2.23). For very short times, imagine a time so short that no collision has occurred, there is only a linear trajectory along which the object is moving. Imagine this is happening with a constant velocity v, then the displacement is linear with t. Hence, by substituting the displacement $\mathbf{x}$ in Eqs. (2.22) and (2.23) with $v \cdot t$, the diffusion constant D will become linear with t and the mean squared displacement (MSD) is proportional to t^2. That regime is known as ballistic diffusion and occurs if a constant drift with velocity v exists beside the normal diffusive motion. Thus, the functional dependence of the MSD offers valuable information about the underlying physical processes. A MSD that is quadratic over time hints to a drift motion whereas a linear dependency indicates a normal diffusive process. If the dependency is neither quadratic nor linear, then anomalous diffusion is present.

2.5 Chirality and Fundamental Symmetries

This thesis is concerned with chiral objects, which are generally defined as those that a lack mirror symmetry. However, as will be discussed in this thesis, in some cases one also needs to consider additional symmetries. The original definition was provided by Lord Kelvin in his Baltimore lectures, where he defined chirality as a geometrical property [18, 19]:

> I call any geometrical figure, or group of points, 'chiral', and say that it has chirality if its image in a plane mirror, ideally realized, cannot be brought to coincide with itself.

The initial and the mirrored object he called *chiroids* but nowadays they're known as *enantiomers*. Chirality in general is an important concept in natural sciences and it occurs across many different disciplines. Most (bio-)molecules are chiral and their enantiomers can thus interact differently with other molecules. This can have dramatic consequences, e.g. the D- and L-enanatiomers of limonene will smell like a lemon or an orange, highlighting the importance of chirality for pharmaceuticals. Also, in nature all natural sugars (including DNA) exist exclusively as right-handed forms. Similarly, almost all naturally occurring amino acids are homochiral (left-handed) and the existance of this homochirality phenomenon relates to questions about the origin of life on Earth. In optics, light can be used to probe and distinguish between two molecuar enantiomers because they exhibit different absorbances for circularly polarized light, as discussed later in Sect. 2.6. Beyond that, there are many more examples in chemistry, material sciences, biology, physics etc. where chirality or enantiomorphism plays an important role.

A helix is an example of a chiral shaped body, as it can exist as a left- or a right-handed enantiomer. In the past, geometric chirality was intuitively assumed to be efficient for propulsion at low Re [4]. Therefore it was always a helical shaped dipolar cork-screw that has been utilized as an artificial micro- or nanopropeller and that has been physically powered with a spinning external magnetic field [7–10, 20, 21]. However, recently a theoretical study has shown that a geometrically achiral shaped body can also be propulsive *via* rotation-translation coupling [15]. The first comprehesnive experimental realization of this concept is demonstrated in Chap. 3.

Mathematically the lack in mirror-symmetry and thus chirality is defined by the symmetry operation of parity:

$$\widehat{P}(x, y, z) \rightarrow (-x, -y, -z). \tag{2.24}$$

The parity operator $\widehat{P}$ inverts any of the three spatial coordinates with respect to an arbitrary origin. If an object changes its sign under parity it is called parity-odd, and is according to Kelvin's definition chiral. If it is invariant under the execution of parity it is parity-even an therefore achiral. Thus, in the case of a parity-even object, its form after applying the parity operator $\widehat{P}$ can be superimposed with its intitial form. Equally the object can be first mirrored on an arbitrary plane and subsequently a rotation about 180° can be found, which superposes the object onto its initial state. For the sake of completeness it should be mentioned that some variables are not directional, and depending on their symmetry under parity they are known as *scalars* ($\widehat{P}-$even) or *pseudoscalars* ($\widehat{P}-$odd). On the contrary, directional variables materialize as ($\widehat{P}-$even) *polar* vectors and ($\widehat{P}-$odd) *axial* vectors, where the latter is also known as a *pseudovector*.

Parity is one of several fundamental symmetries. Another symmetry operation is charge conjugation $\widehat{C}$,

$$\widehat{C}(q) \rightarrow (-q). \tag{2.25}$$

which will transform a particle into it's antiparticle, thus it inverts the sign of a charge and consequently also affects electromagnetic dipole moments. The third fundamental symmetry is time-reversal $\widehat{T}$, which reverses time (e.g. the motion of a particle), or mathematically:

$$\widehat{T}(t) \rightarrow (-t). \tag{2.26}$$

Applying $\widehat{T}$ does not concern the thermodynamic arrow of time, but rather a direction reversal. After applying $\widehat{C}$ or $\widehat{T}$, a property that is invariant or symmetric does not change sign and is called even. On the contrary, a property that changes sign is called odd under the symmetry operation. It is interesting to mention here the CPT-theorem as one of the most fundamental physical symmetrics that is believed to always hold and that is part of modern quantum field theory. It states that the subsequent execution of $\widehat{C}\widehat{P}\widehat{T}$ onto any physical law or quantity has to be invariant.

More than hundred years after the description of chirality given by Lord Kelvin it has been shown by Barron that, if motion is involved, the original definition of chirality has to be extended. The concept of *true* and *false* chirality takes into consideration not only parity $\widehat{P}$, but also the time reversal symmetry $\widehat{T}$ [2]. Barron stated that

> True chirality is exhibited by systems that exist in two distinct enantiomeric states that are interconverted by space inversion (parity $\widehat{P}$), but not by time reversal ($\widehat{T}$) combined with any proper spatial rotation.

The effect of parity and time reversal on a spinning cone is shown in Fig. 2.3a [22]. The stationary cone that spins around its principal axis appears as a chiral object

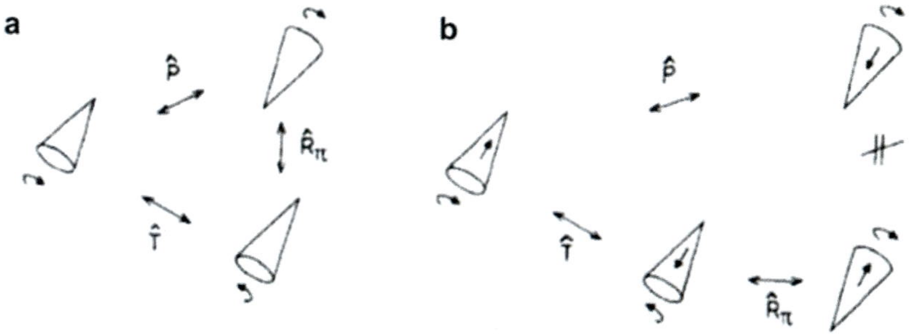

Fig. 2.3 Schematic drawing of true and false chirality of a conical object. **a** A stationary spinning cone is falsely chiral because $\widehat{P} = \widehat{T}\widehat{R}_\pi$. **b** A spinning and simultaneously translating cone is a truly chiral object with $\widehat{P} \neq \widehat{T}\widehat{R}_\pi$. Reprinted from [22] with permission from Elsevier

because it is parity-odd. But application of time-reversal followed by a rotation about 180° (denoted $\widehat{R}_\pi$) will also create the same object, hence it posses *false* chirality. Consequently, the spinning and simultaneously translating cone in Fig. 2.3b exhibits true chirality because it is parity-odd as the time-reversed object it is distinct from the object under the operation of parity.

The existence of true and false chirality for a moving object strongly supports the idea that the symmetry analysis of moving objects, including rotating propellers, and those that possess a dipole moment need to be carefully analyzed. For highly symmetrical dipolar objects, such as a V-shape, one has to take into account not only the transformation properties of the shape but also the transformation of the vector representing the object's dipole moment. Later it will be shown that by changing the orientation of the dipole with respect to the body, an object's symmetry under parity changes, and thus a geometrically achiral shape can be rendered chiral.

2.5.1 Parity and Time-Reversal of Dipole Moments

As mentioned, chirality and other symmetry transformations are not exclusively restricted to geometric shapes. Especially, electromagnetic fields and dipole moments can transform differently and are particularly important for this thesis. Thus, their symmetries are explicitly given here. Interestingly, electric and magnetic quantities obey opposite transformation characteristics upon application of $\widehat{C}$, $\widehat{P}$ and $\widehat{T}$.

An electric dipole moment is created by two charges with opposite sign (q & $-q$) separated by a distance **e**, as can be seen in Fig. 2.4. Therefore, the electric dipole moment $\mathbf{d} = q\mathbf{e}$ and the electric field **E** are time-even polar vectors:

$$\widehat{P}\mathbf{d} = -\mathbf{d}, \quad \widehat{P}\mathbf{E} = -\mathbf{E}, \quad \text{and} \quad \widehat{T}\mathbf{d} = \mathbf{d}, \quad \widehat{T}\mathbf{E} = \mathbf{E}. \tag{2.27}$$

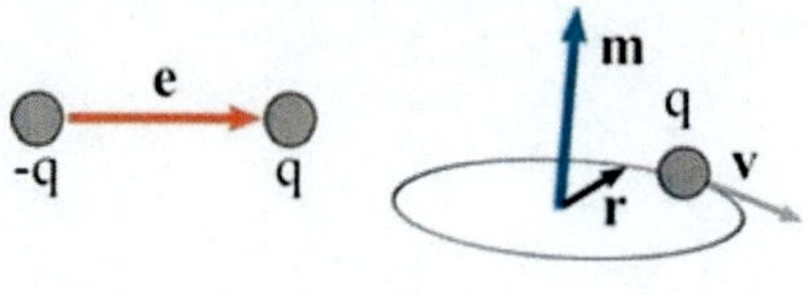

Fig. 2.4 An electric dipole consists of two charges $\pm q$ separated by the vector **e**. A magnetic dipole moment comprises a charge q moving with velocity **v** on a circular trajectory with radius **r** and the cross product yields the dipole moment which is orthogonal to both vectors

Table 2.1 The symmetry properties of an electric dipole moment **d** and and a field **E** as well as a magnetic dipole moment **m** and a field **H** under $\widehat{C}$, $\widehat{P}$, and $\widehat{T}$:

	C	P	T
d & E	−	−	+
m & H	−	+	−

In contrast, a magnetic moment consists of a charge moving on a circular trajectory (Fig. 2.4). This corresponds to a velocity **v** and a distance vector **r**, whilst the cross product between the two constitutes the dipole moment $\mathbf{m} = \frac{q}{2}\mathbf{r} \times \mathbf{v}$. Hence, parity and time-reversal acts on both vectors:

$$\widehat{P}\mathbf{m} = \frac{q}{2}(-\mathbf{r} \times -\mathbf{v}), \quad \text{and} \quad \widehat{T}\mathbf{m} = \frac{q}{2}(\mathbf{r} \times -\mathbf{v}). \tag{2.28}$$

Consequently it follows that the magnetic dipole moment **m** and magnetic field **H** are time-odd axial (pseudo-)vectors:

$$\widehat{P}\mathbf{m} = \mathbf{m}, \quad \widehat{P}\mathbf{H} = \mathbf{H}, \quad \text{and} \quad \widehat{T}\mathbf{m} = -\mathbf{m}, \quad \widehat{T}\mathbf{H} = -\mathbf{H} \tag{2.29}$$

Table 2.1 sums up the transformation properties of the electromagnetic quantities.

2.6 Chiroptical Spectroscopy

Chiral molecules will interact differently with their environment as mentioned above. Either by having different chemical reactivities (with chiral substrate molecules) or by different interaction potentials with (chiral) electromagnetic fields. In classical optical activity experiments that are used to observe chiral molecules, the polarization state of the light as it travels through a liquid sample is changed. The change of the polarization state can then be mapped to the structural properties of the specimen. For instance, the handedness of a chiral solution can be probed by analyzing the response of characteristic molecular and atomic transitions (or in the case of nanostructures the corresponding bands) to circularly polarized light. This thesis is mostly concerned with circular dichroism (CD) spectroscopy, also termed chiroptical spectroscopy.

It detects an absorption difference between left- and right-circularly polarized light, and the related changes in scattering. A brief overview of polarization altering effects is given, before the basis of CD and related effects is explained in greater detail.

Generally, coherent light can exist as linearly polarized (LP) light which is mathematically described by horizontal and transversal (orthogonal) basis vectors and where the electric field vector is oscillating in one plane only. Or it can exists circularly polarized (CP) light that has a 90° phase shift between the two orthogonal polarization states, thus the electric field vector describes a rotation along its forward direction with either left- or right-handed rotation sense, as it propagates, which is known as left-circularly polarized (LCP) and right-circularly polarized (RCP) light, respectively.[1] Any other scenario can be thought of as a superposition of these fundamental polarization states.

Several effects changing the linear and circular polarization states can occur, which are described by the complex refractive index $\tilde{n}$ of a material. It is related to the dielectric function ($\tilde{\epsilon} = \tilde{n}^2$) of a material and thus describes the oscillatory response of the electrons to an external electric field, which can behave differently upon excitation with different polarization states, depending on the symmetry of the molecule or scatterer and its orientation with respect to the light field. The physical origin of this response is the polarizability of the medium, which is in general a (non-isotropic) tensorial quantity. Therefore, in an arbitrary optically anisotropic medium the real and imaginary part of the (complex) refractive index $\tilde{n} = n + i\kappa$ is different for each of the polarization basis vectors and additionally it is wavelength-dependent. The imaginary part of the refractive index κ is called the extinction coefficient and it contains losses due to scattering S and absorption A. A difference in the real part of $\mathfrak{Re}(\tilde{n}) = n$ occurs due to the birefringence, whereas a difference of the imaginary part $\mathfrak{Im}(\tilde{n}) = \kappa$ is known as dichroism. Because the four basic polarization states interact differently with the medium it traverses, there are 4 fundamental effects that can alter the polarization state of light. Those are:

- **Linear Birefringence (LB)**: A medium is *birefringent* if it has different refractive indices for the two orthogonal linear polarizations ($n_{s,p}$ [2]) and therefore the phase velocities and angles of refraction are different by $\Delta n = n_s - n_p$. If an arbitrary polarization is incident it can be projected onto the orthogonal base vectors and one is retarded with respect to the other, hence a phase retardation is introduced. Therefore, a birefringent material can be used as an optical retarder or polarizer.
- **Circular Birefringence (CB)**: Similar to linear birefringence, also the circular polarization states can experience a different refractive index while travelling through a medium. A linearly polarized wave can be decomposed into the basis of a left and right circularly polarized wave with equal amplitude. Thus, if a LP

[1] There is no strict convention in this definition because sometimes RCP is defined as a clock-wise rotation of the electric field when a viewer looks against the propagation direction. Historically, chemists and physicists have used different conventions.

[2] The subscripts s and p stand for the german *senkrecht* (engl. perpendicular) and *parallel* what defines the polarization with respect to the plane of incidence.

wave propagates through a material exhibiting circular birefringence (CB) the polarization plane is rotated. The rotation angle is dependent on the CB given by $\Delta n_{CP} = n_{RCP} - n_{LCP}$. Sometimes this effect is also referred to as circular retardance, in conjunction with Optical Rotation (OR) or simply Optical Activity (OA). Traditionally, CB is measured in a polarimeter. Generally, the CB is only weakly dispersive, but the wavelength dependence of CB is known as optical rotatory dispersion (ORD).
- **Linear Dichroism (LD)**: If the diattenuation or absorbance for the two linear polarization states is different, the material exhibits linear dichroism. It is related to a difference in absorption coefficients: $LD = A_s - A_p \propto \kappa_s - \kappa_p$ and a perfect polarizer would exhibit perfect linear diattenuation. One linear polarization is completely attenuated and vanishes, whereas the orthogonal polarization is unaffected and propagates without losses.
- **Circular Dichroism (CD)**: Light which propagates through a medium exhibiting CD will experience a different absorbance (or attenuation) of left- and right-CP light $CD = A_{LCP} - A_{RCP} \propto \kappa_{LCP} - \kappa_{RCP}$. Because the refractive index is wavelength-dependent, it follows that CD is also dispersive and the characteristic spectral features are a valuable measure of an analytes' structural properties.

The two effects altering the state of circularly polarized light, CB and CD, are especially important for many disciplines in science as they can be used to determine the chirality of an isotropic sample, such as a solution. This is the basis of chiroptical spectroscopy. The chiroptical spectroscopy of single nanoparticles is presented in Chap. 4 of this thesis.

The dichroism and birefringence effects for linear and circular polarizations are Kramers-Kronig [22] related, which implies that if a material shows CB it will necessarily also exhibit CD and vice versa (the same is valid for linear birefringence (LB) and linear dichroism (LD)). A convenient and elegant way to mathematically describe those effects and how they alter the polarization state of light is provided by the Mueller-Stokes formalism (see A.1). The total Mueller matrix of an arbitrary and optically anisotropic sample comprises all of the aforementioned effects [23]:

$$\mathbf{MS}_{\text{sample}} = \begin{bmatrix} 0 & LD & LD' & CD \\ LD & 0 & CB & -LB' \\ LD' & -CB & 0 & LB \\ CD & LB' & -LB & 0 \end{bmatrix}. \tag{2.30}$$

Here the basis is chosen along x and y and the prime (′) denotes the ±45° rotated set of base vectors. In Chap. 4 the Mueller-Stokes formalism is used to understand how experimental imperfections, e.g. in a setup or a sample, can lead to linear polarization effects that mistakenly result in non-zero chiroptical spectra. This is especially important when single-particle measurements are considered and this considerably complicates the spectral interpretation.

Commercial spectrometers provide a rapid and reliable means to measure CD and therefore determine the chirality of an analyte in solution. Usually a cuvette is used to acquire a CD-spectrum of a liquid, which thus contains ensembles of molecules or

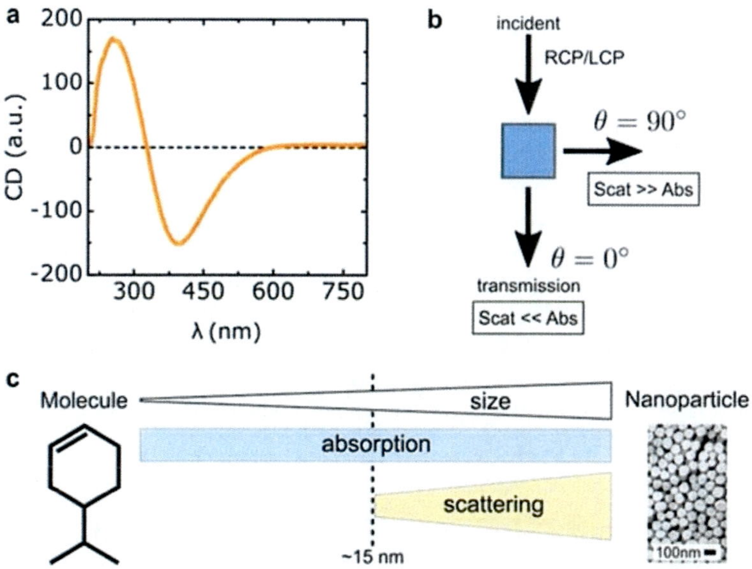

Fig. 2.5 Fundamentals of chiroptical spectroscopy for molecular and artificial colloidal suspensions. **a** Typical CD spectrum of a colloidal dispersion of UV-active Si nanohelices measured in a commercial spectrometer utilizing a cuvette. **b** Schematic drawing of a traditional chiroptical spectrometer utilizing a cuvette and that has the detector mounted in transmission ($\theta = 0°$) or at an angle of $\theta = 90°$ and thus measures absorbance or scattering intensities, respectively. **c** For small objects, e.g. molecules, virtually only absorption is relevant whereas for artificial structures with larger dimensions also scattering is a significant contribution to the occurrence of chiroptical spectra. The onset of scattering for visible wavelengths is around feature sizes of 15 nm

particles. A typical spectrum is shown in Fig. 2.5a. However, a novel approach that uses plasmonic nanostructures to sense chirality has attracted a lot of interest over the past years. It is expected that by coupling the strong electromagnetic fields present in nanostructures with the electromagnetic moments of molecules, the sensitivity of chiroptical measurements can be enhanced by several orders of magnitude [24–26] when compared to the signals recorded with traditional ensemble CD spectrometers. Furthermore, if only one single nanostructure is used for the measurement, then the required sample volume could be dramatically decreased. However, in practice it is not feasible to measure the absorption spectrum of a single nanostructure. But accessing the scattering signal of a single nanostructure is possible, by employing standard dark-field microscopy. Therefore the single-particle approach is almost always based on measuring the circular differential scattering intensity CDSI which is in general defined as [27]:

$$\mathrm{CDSI} = \frac{I_\mathrm{L} - I_\mathrm{R}}{I_\mathrm{L} + I_\mathrm{R}}. \tag{2.31}$$

Here I_R and I_L are the scattering intensities for RCP and LCP light, respectively. As described in Chap. 4, measuring the CDSI of single particles will need a careful analysis as the orientation of the single structure with respect to the input and output beam can influence the signal dramatically. In particular, achiral objects in general also exhibit nonzero CDSI spectra, which can be misinterpreted as an indication of chirality in the sample, even if it is achiral.

It should be emphasized that there is a fundamental difference between traditional chiroptical spectroscopy, this is the (bulk) ensemble CD, which is measured with traditional spectrometers, and the scattering based CDSI which is used to probe single particles. Typically, CD-spectrometers record the differential attenuation of RCP and LCP light propagating through the cuvette containing the sample solution. The absorbances are detected separately under a scattering angle of $\theta = 0°$ with respect to the incident light (see Fig. 2.5b). The spectrometer records intensities (I_{L}, I_{R}), which are described by Lambert–Beer's law. Generally it is not only extinction but also scattering which attenuates the light and CD equals the normalized differential intensity [28]:

$$\mathrm{CD} = \frac{I_{\mathrm{L}} - I_{\mathrm{R}}}{I_{\mathrm{L}} + I_{\mathrm{R}}} = \frac{-2.303\,(\varepsilon_{\mathrm{L}} - \varepsilon_{\mathrm{R}})\,c\ell}{2} + \frac{\sigma_{\mathrm{L}}(0°) - \sigma_{\mathrm{R}}(0°)}{2r^2 + \sigma_{\mathrm{L}}(0°) + \sigma_{\mathrm{R}}(0°)}, \tag{2.32}$$

with c the concentration, ℓ the propagation length in the sample, r the distance between sample and detector, and $\sigma(\theta = 0°)$ the scattering cross-section of the sample. A difference in extinction $\Delta\varepsilon = \varepsilon_{\mathrm{L}} - \varepsilon_{\mathrm{R}}$ contains the losses due to absorption ($A_{\mathrm{R,L}}$) as well as scattering ($S_{\mathrm{R,L}}$) in the sample, and hence $\Delta\varepsilon = (A_{\mathrm{L}} - A_{\mathrm{R}}) + (S_{\mathrm{L}} - S_{\mathrm{R}})$. Because molecules are small compared to the wavelength of visible light, the measurements occur in the Rayleigh limit and contributions due to scattering vanish. Consequently, $S_{L,R} \approx 0$ and $\sigma(0°) \approx 0$ in Eq. (2.32) and in practice only absorption is significant for the attenuation. Hence CD is typically defined as a difference of absorbances of LCP and RCP light only:

$$\mathrm{CD} = \frac{I_{\mathrm{L}} - I_{\mathrm{R}}}{I_{\mathrm{L}} + I_{\mathrm{R}}} \simeq \frac{-2.303\,(\varepsilon_{\mathrm{L}} - \varepsilon_{\mathrm{R}})\,c\ell}{2} \propto A_{\mathrm{L}} - A_{\mathrm{R}}. \tag{2.33}$$

On the other hand, artificial nanostructures are hundreds of nanometers in size and scattering may not be neglected anymore. In fact, scattering of artificial nanostructures plays an important role for the emergence of the CD signal if the characteristic length d of the scatterers is $d > \frac{\lambda}{20}$ [28] as schematically depicted in Fig. 2.5c. Then, propagation through a suspension of nanostructures causes losses due to both, absorption and scattering, and the acquired CD-spectrum depends on the setup geometry, i.e. the distance r.

More importantly, the scattering angle θ then influences the differential signal. This implies, if the intensities are detected at any other angle except zero, $\theta \neq 0°$, scattering dominates the recorded intensities, whereas the impact of absorption vanishes ($A_{L,R} \approx 0$). Therefore, a spectrum acquired at $\theta \neq 0°$ has a Circular Differential Scattering Intensity (CDSI) given by [28]:

$$\mathrm{CDSI} = \frac{I_\mathrm{L} - I_\mathrm{R}}{I_\mathrm{L} + I_\mathrm{R}} = \frac{-2.303\,(\varepsilon_\mathrm{L} - \varepsilon_\mathrm{R})\,c\ell}{2}\left[\frac{(1+\cos\theta)^2}{2\left(1+\cos^2\theta\right)}\right] + \frac{\sigma_\mathrm{L}(\theta) - \sigma_\mathrm{R}(\theta)}{\sigma_\mathrm{L}(\theta) + \sigma_\mathrm{R}(\theta)}. \tag{2.34}$$

A CDSI can thus not only be measured for a single nanoparticle, but rather also on an ensemble in a traditional CD-spectrometer with the detector mounted at an angle other than $\theta = 0°$ (Fig. 2.5c). Another important implication of Eqs. (2.33) and (2.34) is that absorption based CD ($\theta = 0°$) and scattering based CDSI ($\theta \neq 0°$) do not necessarily yield the same spectra and sign.

Measurements of colloidal ensembles conducted with a commercial spectrometer that had its detector mounted at $\theta = 0°$ and $\theta = 90°$ are shown in Sect. 4.3. The single-particle spectroscopy setup described in Sect. 4.2 records a CDSI, because it is based on dark-field microscopy. However, single-particle measurements and corresponding colloidal ensemble CDSI spectra acquired with the traditional spectrometer at $\theta = 90°$ are still in general different quantities. The latter records light at a single scattering angle whereas the novel setup uses a large range of different scattering angles fixed by the Numerical Aperture (NA) of the condenser.

2.7 Physical Vapour Deposition of Micro- and Nanostructures

Physical Vapor Deposition (PVD) methods are established in research as well as industry for thin film coating of substrates. A source material is heated and evaporated either thermally or by using an electron-beam. This will transfer the atoms and molecules into the gas phase where they undergo ballistic motion. Because vacuum conditions are used in the evaporation chamber, they will not be deflected due to a lack of potential collision partners. Thus, a directed flux of the source material is generated and is usually incident onto a substrate which has its surface perpendicular to the flux direction. The atoms will arrive on the substrate and after some temperature-dependent diffusion they will be physically adsorbed. Because the atoms tend to adsorb in an energetically favorable position, normally a thin film will grow on the substrate. If the substrate is crystalline and the lattice constant of the incident atoms does not differ too much from the substrates' lattice constant, an epitaxial layer can be grown on top. This is a layer for which at least one crystal axis of the substrate is conserved during growth. Otherwise, amorphous growth will occur leading to a film that is not perfectly flat. This leads to nucleation centers that (randomly) build up on the substrate. In this case, a geometric shadowing effect occurs under non-perpendicular incident angles of the vapour flux. The evaporation technique that uses shallow incident angles together with substrate rotation is known as Glancing Angle Deposition (GLAD) or sometimes oblique angle deposition (in the absence of rotation). Micro- and nanostructures grown by utilizing well-controlled incident angles are sometimes denoted as sculptured thin films.

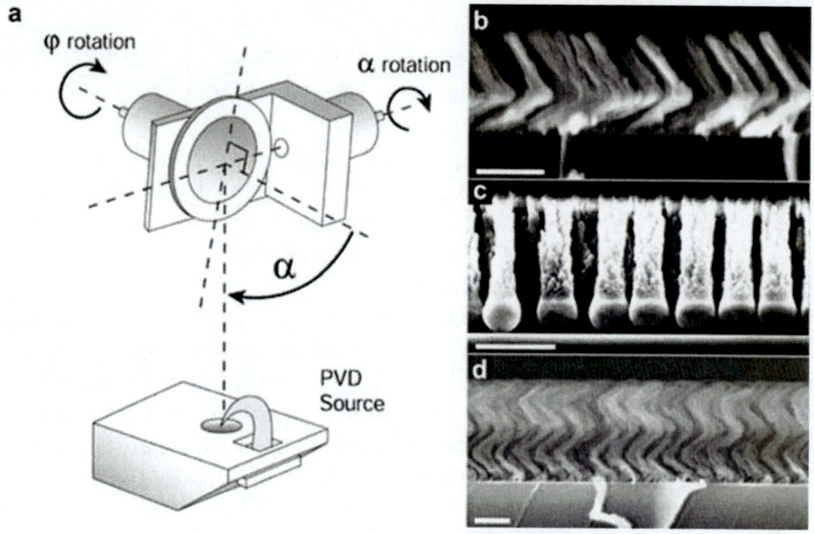

Fig. 2.6 Glancing Angle Deposition (GLAD) of micro- and nanostructures. **a** Schematic drawing of the GLAD setup, where α is the angle between the substrate normal and the flux direction, whereas φ describes the azimuthal rotation. Reprinted with permission from [29]. Copyright 2003, American Vacuum Society. **b** Scanning Electron Microscopy (SEM) image showing that a tilted incidence ($\alpha < 90°$) on a substrate pre-patterned with Au nondots yield a tilted (Ni) column and by changing $\varphi \rightarrow \varphi + 180°$ before the subsequent growth of a second column (SiO_2), a zigzag structure forms (scale bar 200 nm). **c** SEM image of microrods perpendicular to the substrate, here TiO_2 pillars on top of 500 nm diameter silica beads, that were grown with a fast substrate rotation (scale bar 1 μm). **d** The SEM reveals that by slowly and constantly varying φ during a GLAD deposition, helicoidal structures will form, here left-handed SiO_2 helices on Au nondots (scale bar 200 nm)

A comprehensive overview of GLAD is found in Ref. [3] and a schematic drawing of the standard setup is shown in Fig. 2.6a, where the incident angle of the vapour flux (α) is measured with respect to the surface normal of the substrate. The already mentioned shadow effect leads to the growth of columnar structures that are tilted with respect to the substrate surface. The freedom to rotate the substrate around the surface normal by an angle φ provides control over the orientation of columns, and even more importantly their shape can be altered too. A growth process without rotation will lead to a tilted columnar growth on the surface, whereas a change of φ by 180° between the evaporation of two tilted columns forms a zigzag or V-shape (Fig. 2.6a). A continuous (fast) rotation of φ results in a perpendicular column (Fig. 2.6b). If the rotation speed of φ is slower and constant[3] a helicoidal structure will form as can be seen in Fig. 2.6c. Thus the GLAD technique is a valuable tool to fabricate nano- and microstructures with complex three-dimensional shapes that have lowest symmetry, (e.g. the C_1 symmetry of a helix with $n \notin \mathbb{N}$ turns). Moreover, a wide choice of materials (oxides, metals, organics) can be evaporated. Especially important is the possibility to alloy materials by simultaneously evaporating two or more source

[3] The rotation has to be constant with respect to the amount of incident material, thus it is permanently adjusted if the evaporation rate changes.

materials. The method is "wafer-scale", giving the opportunity to fabricate up to 10^{11} nanostructures on one two inch wafer during one deposition process. After the growth process, GLAD structures can be removed from the substrate via ultrasonication and transferred to a liquid. Thereby colloidal solutions with concentrations up to tens of picomolar are formed and sometimes sacrificial layers, e.g. NaCl, are useful to facilitate the removal of structures form the substrate.

As part of this thesis work, micro- and nanostructures have been grown with an extended GLAD technique [30]. Usually the shadowing happens randomly on the substrate but here it was facilitated using pre-patterned substrates with a defined nucleation seed template. For this, a (hexagonally) close-packed monolayer of silica beads, prepared with a Langmuir–Blodgett-trough can be used [31, 32] (see Fig. 2.6b). In contrast, the GLAD structures in Fig. 2.6a and c utilize non-close-packed monolayers of gold nanodots with only a few nanometers in diameter. They have been prepared with block-copolymer micellar nanolithography [33]. Additionally, substrate cooling was sometimes employed during evaporation to reduce the energy the adatoms have after they get adsorbed on the substrate. Generally evaporants have a high surface mobility due to their thermal energy and hence they have the tendency to diffuse and form highly symmetric equilibrium shapes. This is particularly problematic for noble metals, whereas the growth of complex-shaped nano- and microstructures is facilitated by lowering the surface temperature. However, GLAD can obviously also be employed to fabricate very simple patchy particles. For this a thin film of a few tens of nanometers can be coated onto a micron sized bead in order to create a spherical particle with two distinct materials on its hemispheres, also known as a Janus particle (JP). These growth procedures are employed to fabricate the particles that are examined in the subsequent chapters.

References

1. Happel J, Brenner H (1983) Low Reynolds number hydrodynamics with special applications to particulate media. Monographs and textbooks on mechanics of solids and fluids, 1st paperback edn. Martinus Nijhoff, The Hague
2. Barron LD (1983) Molecular light scattering and optical activity. Cambridge University Press, Cambridge
3. Hawkeye MM, Taschuk MT, Brett MJ (2014) Glancing angle deposition of thin films: engineering the nanoscale. Wiley, Hoboken
4. Purcell EM (1977) Life at low Reynolds number. Am J Phys 45:3
5. Qiu T, Lee T-C, Mark AG, Morozov KI, Muenster R, Mierka O, Turek S, Leshansky AM, Fischer P (2014) Swimming by reciprocal motion at low Reynolds number. Nat Commun 5
6. Baranova NB, Zel'dovich BY (1978) Separation of mirror isomeric molecules by radio-frequency electric field of rotating polarization. Chem Phys Lett 57:435
7. Zhang L, Abbott JJ, Dong L, Kratochvil BE, Bell D, Nelson BJ (2009) Artificial bacterial flagella: fabrication and magnetic control. Appl Phys Lett 94(6):064107
8. Ghosh A, Fischer P (2009) Controlled propulsion of artificial magnetic nanostructured propellers. Nano Lett 9(6):2243–2245
9. Schamel D, Pfeifer M, Gibbs JG, Miksch B, Mark AG, Fischer P (2013) Chiral colloidal molecules and observation of the propeller effect. J Am Chem Soc 135(33):12353–12359

10. Schamel D, Mark AG, Gibbs JG, Miksch C, Morozov KI, Leshansky AM, Fischer P (2014) Nanopropellers and their actuation in complex viscoelastic media. ACS Nano 8(9):8794–8801
11. Wu Z, Troll J, Jeong H-H, Wei Q, Stang M, Ziemssen F, Wang Z, Dong M, Schnichels S, Qiu T, Fischer P (2018) A swarm of slippery micropropellers penetrates the vitreous body of the eye. Sci Adv 4(11):eaat4388
12. Palagi S, Mark AG, Reigh SY, Melde K, Qiu T, Zeng H, Parmeggiani C, Martella D, Sanchez-Castillo A, Kapernaum N, Giesselmann F, Wiersma DS, Lauga E, Fischer P (2016) Structured light enables biomimetic swimming and versatile locomotion of photoresponsive soft microrobots. Nat Mater 15(6):647–653
13. Doi M, Makino M (2016) Separation of chiral particles in a rotating electric field. Phys Fluids 28(9):093302
14. Kim SA, Karrila SJ (1991) Microhydrodynamics: principles and selected applications. Butterworth-Heinemann series in chemical engineering. Butterworth-Heinemann
15. Morozov KI, Mirzae Y, Kenneth O, Leshansky AM (2017) Dynamics of arbitrary shaped propellers driven by a rotating magnetic field. Phys Rev Fluids 2(4):044202
16. Koenig SH (1975) Brownian motion of an ellipsoid. A correction to Perrin's results. Biopolym: Orig Res Biomol 14(11):2421–2423
17. Kuipers BWM, Van de Ven MCA, Baars RJ, Philipse AP (2012) Simultaneous measurement of rotational and translational diffusion of anisotropic colloids with a new integrated setup for fluorescence recovery after photobleaching. J Phys: Condens Matter 24(24):245101
18. Kelvin L (1894) The molecular tactics of a crystal. Clarendon Press, Oxford
19. Kelvin L (1904) Baltimore lectures on molecular dynamics and the wave theory of light. CJ Clay & Sons
20. Bell DJ, Leutenegger S, Hammar KM, Dong LX, Nelson BJ (2007) Flagella-like propulsion for microrobots using a nanocoil and a rotating electromagnetic field. In: Proceedings 2007 IEEE international conference on robotics and automation. IEEE, , pp 1128–1133
21. Smith EJ, Makarov D, Sanchez S, Fomin VM, Schmidt OG (2011) Magnetic microhelix coil structures. Phys Rev Lett 107(9):097204
22. Barron LD (1986) True and false chirality and parity violation. Chem Phys Lett 123(5):423–427
23. Arteaga O, Kahr B (2013) Characterization of homogenous depolarizing media based on mueller matrix differential decomposition. Opt Lett 38(7):1134–1136
24. Hendry E, Carpy T, Johnston J, Popland M, Mikhaylovskiy RV, Lapthorn AJ, Kelly SM, Barron LD, Gadegaard N, Kadodwala M (2010) Ultrasensitive detection and characterization of biomolecules using superchiral fields. Nat Nanotechnol 5(11):783–787
25. Zhao Y, Askarpour AN, Sun L, Shi J, Li X, Alù A (2017) Chirality detection of enantiomers using twisted optical metamaterials. Nat Commun 8(1):1–8
26. Collins JT, Kuppe C, Hooper DC, Sibilia C, Centini M, Valev VK (2017) Chirality and chiroptical effects in metal nanostructures: fundamentals and current trends. Adv Opt Mater 5(16):1700182
27. Bustamante C, Maestre MF, Tinoco I (1980) Circular intensity differential scattering of light by helical structures. i. theory. J Chem Phys 73(9):4273–4281
28. Bustamante C, Tinoco I, Maestre MF (1983) Circular differential scattering can be an important part of the circular dichroism of macromolecules. Proc Natl Acad Sci 80(12):3568–3572
29. Dick B, Brett MJ, Smy T (2003) Controlled growth of periodic pillars by glancing angle deposition. J Vac Sci Technol B: Microelectron Nanometer Struct Process, Meas, Phenom 21(1):23–28
30. Mark AG, Gibbs JG, Lee T-C, Fischer P (2013) Hybrid nanocolloids with programmed three-dimensional shape and material composition. Nat Mater 12(9):802–807
31. Langmuir I, Blodgett KB (1935) Über einige neue methoden zur untersuchung von monomolekularen filmen. Kolloid-Zeitschrift 73(3):257–263
32. Reculusa S, Ravaine S (2003) Synthesis of colloidal crystals of controllable thickness through the langmuir- blodgett technique. Chem Mater 15(2):598–605
33. Glass R, Möller M, Spatz JP (2003) Block copolymer micelle nanolithography. Nanotechnology 14(10):1153

Chapter 3
Motion of Chiral and Achiral Structures at Low Re

This chapter is based on and contains excerpts and figures from the articles "Role of symmetry in driven propulsion at low Reynolds number" [1] and "Characterization of active matter in dense suspensions with heterodyne laser Doppler velocimetry" [2]. Contributions of coauthors are indicated.

3.1 General Motivation

Physically powered micro- and nanostructures have been proposed for their use in biomedical applications, such as targeted drug delivery or minimally invasive surgery [3]. They are small and are intended to be employed in biological tissues, that are Non-Newtonian and generally highly viscous. Consequently they operate at low Re environments, which puts severe constraints on their propulsion mechanism due to the time-reversibility of the Stokes equation (see Eq. (2.5) and Sect. 2.2). To overcome this, Nobel laureate E. Purcell once famously remarked in his lecture "Life at low Reynolds number" [4]:

> Turn anything - if it isn't perfectly symmetrical, you'll swim.

Nevertheless, the design of artificial microbots has always been inspired by nature, which developed some highly sophisticated mechanisms in order to generate locomotion of small micro-organisms at low Re. One example are cells that use flexible cilia on their surface and beat them in a coordinated, wave-like patterns [5]. Spermatozoa use beating flexible flagella in order to move [6]. However, for artificial structures a synchronized and coordinated complex motion of individual body parts is a non-trivial task. Also the miniaturization of flexible and at the same time stable materials remains challenging, although there has been some experimental realizations for that [7].

J. Sachs, *Motion, Symmetry & Spectroscopy of Chiral Nanostructures*, Springer Theses,
https://doi.org/10.1007/978-3-030-88689-9_3

Another particularly intriguing example are thus bacteria, e.g. *E. coli*, which spin their helical flagella [8] which causes them to move in low Re conditions. This approach is simpler to realize experimentally, as it allows one to exploit the chirality of a rigid corkscrew-like shape together with rotation-translation coupling. One should note, that spinning a rigid structure is not identical to the swimming bacterium, as the bacterial cell conserves torque and therefore its body spins in the opposite direction. Spinning a rigid structure with a rotating field imparts momentum to the fluid. Nevertheless, this approach is attractive, as it does not require any special materials to realize flexible filaments. The propulsion of artificial microhelices has been first demonstrated by fabricating helical shaped microstructures with a PVD method that has been combined with photolithography to incorporate a magnetic moment [9]. A rotating external magnetic field was then used to align the microhelix constantly, and thus causing it to rotate around its long axis. Consequently, this led to a translational motion along this axis, similar to a corkscrew. However, the used structures remained comparably big ($\geqslant 40\,\mu$m) and their motion was generally not well-controlled. Shortly thereafter, a much smaller magnetic microscrew that was only two microns long has been fabricated with GLAD [10]. In addition to the previous report, the microhelix could be steered by turning the rotation plane of the external field. Thereby it reliably proved the concept of magnetically actuated microscrews. Subsequently, the optimal length of the same micro- and nanohelices [11] and their actuation in Non-Newtonian fluids [12] has been investigated. Just recently it was verified that they can indeed move through more complex biological tissue like the vitreous humor of a porcine eye [13], pushing them closer towards possible biomedical applications. Beside rotating magnetic fields, also gradient or oscillating magnetic fields can be utilized to move magnetic microstructures through viscous fluids. However, this goes beyond the scope of this thesis and more details on other propulsion mechanisms can be found elsewhere [14].

Due to the inspiration from biology many experiments exploit the hydrodynamic rotation-translation coupling inherent to chiral corkscrew-like shapes. It was thus widely taken for granted that chirality is a prerequisite for efficient propulsion by means of an external rotating field, and a helix/corkscrew was considered to be the optimal shape for converting a rotation into linear motion. On the other hand, the question whether much simpler shapes can also exhibit efficient propulsion *via* rotation-translation coupling at low Re has, surprisingly, not been answered. Simpler shapes are beneficial, because it is generally less complicated to fabricate micro- and nanoscale objects with high spatial symmetry rather than 3-dimensional helical shapes with low symmetry. Non-helical, but complex chiral shapes have been investigated for propulsion by using randomly assembled and thus arbitrary shaped magnetic clusters [15, 16]. It was demonstrated that they can still move with high translational velocities upon rotation-translation coupling but still the shapes are complex and cannot be easily reproduced. Beside the random shapes, another study reported about locomotion of a cluster of three ferromagnetic beads that were chemically bound to each other exhibiting high spatial symmetry. The propulsion characteristics of the three-bead cluster was examined and since three spheres cannot be arranged to show 3D spatial chirality, the study concluded that seemingly achiral objects are propulsive [17]. However, the authors did not incorporate the transformation properties of the dipole moment into their analysis, which is insufficient for a complete description of the objects' symmetry [18]. In this thesis it is shown that a full description is only possible, if the symmetry analysis also accounts for the dipole moment of the dipolar shape. In fact, the attached dipole moment renders the 3-bead cluster chiral, which had not been noticed by the authors. Therefore, no achiral object had thus far been propelled by rotation-translation coupling—something that is realized in this thesis. Furthermore, the role of the orientation of the dipole moment with respect to the body axes is closely examined and shown to be decisive whether an object propels or not.

A recently published theoretical study has presented an analytical solution of the locomotion of an arbitrary shaped and magnetized object, which is spun by an external magnetic field [19]. Thereby the authors revealed the connection between the propulsion velocity and the geometry of the shape, as well as the orientation of the magnetic dipole moment. The theory demonstrates, that geometric chirality of the shape is not necessary for an object to translate due to a rotating magnetic field. While it was already known that a simple and highly symmetrical V-shaped object, whose symmetry is identical to the aforementioned 3-bead cluster, exhibits non-zero elements in the coupling tensor (see Eq. (2.14) in Sect. 2.13), it was now shown, that only for certain orientations of the magnetic dipole moment does a rotation lead to translation. Depending on the dipoles' orientation the authors predicted either non-, uni- or bidirectional propulsion characteristics for the V-shape. Additionally, simulations determined that if a propellers shape is geometrically achiral, then an arc-segment is the optimal design and can display velocities comparable to those realized with corkscrew propellers. However, the V-shape is just an approximation of the arc-segment and consequently the symmetries of both shapes are identical and both are suitable for an experimental demonstration.

In this chapter, the theoretical predictions about the non-, bi- or unidirectional propulsion characteristics of V-shaped objects depending on the orientation of the dipole moment [19] were experimentally examined. First in experiments with macroscopic arc-shaped propellers that permitted the magnetic moment to be precisely oriented with respect to the body axes, and subsequently with V-shaped microcstructures that permitted the use of electric fields and electric moments as well. All experiments were in low Re environments. It is shown that the geometric shape alone is not sufficient to determine an objects' chirality, if it is dipolar. It is also proven that propulsive magnetic objects, like the aforementioned 3-bead cluster, are necessarily chiral, and thereby the claim of a previous report of a propelling achiral structure is shown to be incorrect. Ultimately, the first experimental demonstration of a truly achiral object was realized during this thesis, but this required the actuation by means of an electric field. The experiments and findings validate that geometric chirality is not the property that determines the propulsion characteristics of objects exploiting rotation-translation coupling. Therefore, a rigorous symmetry analysis was developed to explain and predict the behaviour of the V-shape propellers, which includes not only parity ($\widehat{P}$), but also charge conjugation ($\widehat{C}$). The theoretical framework and the symmetry analysis has been derived in collaboration with Alexander

M. Leshanksy and his colleagues Konstantin I. Morozov and Oded Kenneth from the *Technion - Israel Institute of Technology, Haifa*. Experiments with macroscopic propellers were conducted with the help of Tian Qiu, whereas part of the microscale experiments were performed by Nico Segreto under my supervision, both part of the *Micro-, Nano- and Molecular Systems* group at the *Max Planck Institute for Intelligent Systems, Stuttgart*.

After unraveling the motion of highly symmetrical V-shaped objects propelling due to rotation-translation coupling, it was interesting to examine propulsion of even simpler shapes. An simpler shape needs to exhibit an even higher symmetry than the "V" shaped particle, which has two symmetry planes. For instance, a cube, an ellipsoid or a sphere which has three symmetry planes and which clearly is achiral. For those shapes rotation-translation coupling at low Re fails and consequently physically powered propulsion is impossible. However, an artificial model system is a spherical JP, which posses different chemical reactivities on its two hemispheres (and thus has one symmetry-direction broken) to generate active locomotion. While microscopic colloidal Janus swimmers are well-established, it is to date not clear if this means of propulsion is effective for nanoswimmers. Experimental data could bring clarification but is, so far lacking. Thus, the propulsion characteristics of nanoscale Janus particles was also investigated in this thesis. Because their sizes were too small to employ classical bright-field microscopy they have been examined with light scattering techniques, which rely on measuring particle ensembles with high densities, e.g. DLS, DDM and SH-LDV. The latter has been identified as the most suitable technique for studying the motion in Janus particle ensembles and first preliminary results on the size-dependent propulsion characteristics are presented. The Laser-Doppler-Velocimetry experiments were conducted in Thomas Palberg's laboratory at the *Institute of Physics, Johannes-Gutenberg University Mainz* in collaboration with Denis Botin who set up the basic optics and conducted the experiments as well as Sai Nikhilesh Kottapalli from the *Micro-, Nano- and Molecular Systems* group at the *Max Planck Institute for Intelligent Systems, Stuttgart* who helped with the fabrication of the Janus particles.

3.2 Theoretical Background

The fundamentals explaining the motion of micro- or nanostructures in a fluidic environment at low Re are given in Chap. 2, whereas the governing equations for an object that couples its rotation to its translation are found in Sect. 2.3. The general theoretical framework that describes an arbitrary magnetized body, e.g. a shape like an arc or a "V", that is driven by an external torque is explained elsewhere [19]. An application of this theory to the shapes that have been experimentally implemented during this thesis is also presented in literature [1]. The full derivation is, however, beyond the scope of this thesis. Thus, this section will merely motivate the mathematical background and recap the most important findings in order to give an intuitive understanding of the propulsion these shapes exhibit upon rotation-translation coupling at low Re. Note, that the geometrical symmetry an arc and a V-shape, as well as a 3-bead cluster, exhibit are identical. All three shapes posses two mutually perpendicular planes of reflection symmetry, and hence the presented arguments are valid for any of these structures.

In Sect. 2.3 (Eq. (2.12)) it was shown that in the absence of external forces, the translational and rotational velocity of an object are both linearly dependent on the torque exerted on it ($\boldsymbol{U} = \mathcal{G} \cdot \boldsymbol{L}, \quad \boldsymbol{\Omega} = \mathcal{F} \cdot \boldsymbol{L}$). They are related by the coupling and rotation mobility tensors $\mathcal{G}$ and $\mathcal{F}$. Under the assumption that the external torque $\boldsymbol{L}$ is exerted because of the interaction of a magnetic dipole $\boldsymbol{m}$ with a magnetic field $\boldsymbol{H}$, the relationship is:

$$\boldsymbol{L} = \boldsymbol{m} \times \boldsymbol{H}. \tag{3.1}$$

By continuously rotating the external field vector $\boldsymbol{H}$ the dipole would constantly align and therefore continuously rotate an object to which it is fixed. Note that the object's rotation axis depends on how the dipole moment is attached onto the body. The rotation vector $\boldsymbol{L}$ (torque) will always be perpendicular to $\boldsymbol{m}$ and thus the orientation of $\boldsymbol{m}$ with respect to the object's body axes is important. Consequently, for an arbitrary alignment of the magnetic dipole, the object will rotate around an arbitrary axis (but still perpendicular to $\boldsymbol{m}$). This highlights the importance to consider not only the shape, but also the dipole moment for a rigorous symmetry analysis of a propeller driven by means of an external torque via rotation-translation coupling. As already mentioned, this fact has been overlooked in recently published work that claimed a seemingly "achiral" object exhibits propulsion by rotation-translation coupling [17].

By combining the two linear equations from Eq. (2.12) the translational velocity due to an exerted torque can be expressed as:

$$\boldsymbol{U} = \mathcal{G} \cdot \mathcal{F}^{-1} \cdot \boldsymbol{\Omega}. \tag{3.2}$$

For convenience, the principal axes of $\mathcal{F}$ define the body-coordinate-frame. This is a coordinate system which is rigidly affixed to a body and here it is retrieved by diagonalizing the rotation mobility tensor $\mathcal{F}$.[1] A magnetic dipole moment $\boldsymbol{m}$, that is attached to the body, has thus also fixed coordinates in the body-coordinate-frame. In contrast, for a rotating external magnetic field it is more convenient to mathematically express it in a second, external coordinate system, called the lab-coordinate-frame. A constant rotation in the xy in lab coordinates then reads $\boldsymbol{H} = H(\cos(\omega t), \sin(\omega t), 0)$.

In Fig. 3.1a the three principal axes of rotation $\boldsymbol{e}_1, \boldsymbol{e}_2, \boldsymbol{e}_3$, that constitute its body-coordinate-frame, are visualized for an arc-shaped body. Assuming the arc has a magnetic dipole moment that causes it to reorient due to an external magnetic field, the built-in body-coordinate-system is obviously also reorienting its alignment with respect to the lab-coordinate-frame. Mathematically both coordinate systems are

[1] Many textbooks, e.g. see [20], define the principal axes of either rotation, translation or coupling tensors based on eigenvectors that diagonalize the corresponding hydrodynamic resistance tensors, whereas here its is defined in terms of the hydrodynamic mobility, which is the inverse of the resistance tensors (see (2.10)). For arbitrary shapes the two definitions yield different frames.

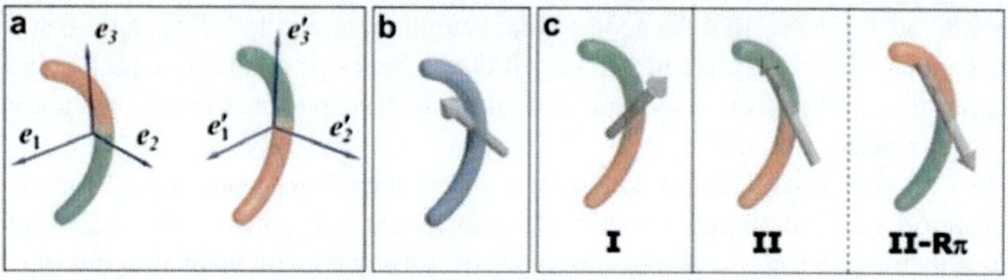

Fig. 3.1 Schematic drawing for an object with mutually orthogonal planes of reflection symmetry, e.g. a V-shape or a 3-bead cluster, here depicted for an arc. **a** The three principal axes of rotation $\boldsymbol{e}_1$, $\boldsymbol{e}_2$ and $\boldsymbol{e}_3$, which are identical to the body-coordinate-frame, yield an arbitrary choice towards which end the $\boldsymbol{e}_3$ axis is pointing. **b** By attaching a dipole moment that lies in the $\boldsymbol{e}_1\boldsymbol{e}_3$-plane it is indistinguishable towards which end $\boldsymbol{m}$ is pointing. **c** If $\boldsymbol{m}$ is tilted out-of-plane, a component along $+\boldsymbol{e}_2$ or $-\boldsymbol{e}_2$ results in different objects. It can now be distinguished if the moment points to the left or the right side of the arc, which is clear if the second object is rotated by 180° around $\boldsymbol{e}_1$

related to each other by a rotation $\boldsymbol{R}$ that can be parameterized by the three Euler angles (φ, θ, ψ), describing a $z - x - z$ rotation. Under the assumption that the magnetic field rotates in the xy and that the magnetic moment $\boldsymbol{m}$ turns in-sync with the field, it is rotating about the z-axis with angular velocity $\boldsymbol{\Omega} = \omega\hat{z}$. The arc's propulsion velocity (Eq. (3.2)) can then be projected onto this vector [19],

$$\frac{U_z}{\omega\ell} = \widehat{\boldsymbol{\Omega}} \cdot \mathfrak{C} \cdot \widehat{\boldsymbol{\Omega}}, \tag{3.3}$$

where $\mathfrak{C} = \mathscr{G} \cdot \mathscr{F}^{-1}$ is the dimensionless *chirality matrix*,[2] ℓ is the characteristic waist of the arc and $\widehat{\boldsymbol{\Omega}}$ is the unit angular velocity vector. Notice that U_z in Eq. (3.3) specifies the velocity in the lab-coordinate-frame, but $\mathfrak{C}$ is fixed in the body-coordinate-frame, in which $\widehat{\boldsymbol{\Omega}}$ becomes:

$$\widehat{\boldsymbol{\Omega}} = \boldsymbol{\Omega}/\omega = \hat{z} = \sin(\theta)\sin(\psi)\boldsymbol{e}_1 + \sin(\theta)\cos(\psi)\boldsymbol{e}_2 + \cos(\theta)\boldsymbol{e}_3. \tag{3.4}$$

The arc, or V-shape, possesses only two nonzero elements in its rotation-translation coupling tensor $\boldsymbol{C} = \mathscr{G}^{-1}$ (see Eq. (2.14)). Consequently, the chirality matrix $\mathfrak{C}$ will also have only two nonzero entries and for an arc with equal arms it follows that $\mathfrak{C}_{23} = \mathfrak{C}_{32}$, and hence Eq. (3.3) reduces to [19]:

$$\frac{U_z}{\omega\ell} = \mathfrak{C}_{23} \cdot \cos(\psi)\sin(2\theta). \tag{3.5}$$

Here $\mathfrak{C}_{23} = \frac{1}{2\ell}\mathscr{G}_{23}({\mathscr{F}_2}^{-1} + {\mathscr{F}_3}^{-1})$ with $\mathscr{G}_{23} = \mathrm{C}_{23}^{-1}$ is the only non-zero entry left in the rotation-translation coupling mobility tensor $\mathscr{G}$ and $\mathscr{F}_2, \mathscr{F}_3$ are the two major

[2] The entries of $\mathfrak{C}$ are determined by $\mathfrak{C}_i \equiv \mathscr{G}_{ii}/(\mathscr{F}_i\ell)$ on the diagonal and $\mathfrak{C}_{ij} = \frac{1}{2\ell}\left(\mathscr{G}_{ij}/\mathscr{F}_j + \mathscr{G}_{ji}/\mathscr{F}_i\right)$ with $i \neq j$ and no summation over repeating indices for off-diagonal elements.

eigenvalues of $\boldsymbol{\mathcal{F}}$ ($\mathcal{F}_1 \leq \mathcal{F}_2 \leq \mathcal{F}_3$). The Euler angle θ defines a precession angle between the axis $\boldsymbol{e}_3$ of the body-coordinate-frame and the $\hat{z}$-axis in lab coordinates, around which the external field rotates. This has important consequences as Eq. (3.5) implies that if the object's $\boldsymbol{e}_3$ axis is not precessing ($\theta = 0°$) then $U_z = 0$ and the arc is not translating.

Generally, a stable rotation of a body only occurs for rotations around the principal axes of rotation because the rotation mobility tensor is then diagonal. Here the principal axes are identical with the body-coordinate axes and hence a rotation of the arc around $\boldsymbol{e}_1$, $\boldsymbol{e}_2$ or $\boldsymbol{e}_3$ will not lead to a precession, and thus no propulsion. It can be mathematically proven that if the magnetic moment is aligned parallel to one of the principal rotation axes $\boldsymbol{m} \parallel \boldsymbol{e}_i$, then its rotational motion must proceed around one of the other principal axes, and therefore $U_z = 0$ [1]. In turn, if the arc is spun around any axis that is not a principal axis of rotation, it will precess. For slow rotation frequencies of the external magnetic field the precession is still negligible and thus no considerable net translation is expected. If the field rotates with higher frequencies, the precession becomes more pronounced and hence the translational velocity is also increasing. The precession onset is called tumbling-to-wobbling transition and happens at $\omega_{\mathrm{t-w}}$. Beyond a certain frequency, at the so called step-out frequency ($\omega_{\mathrm{s-o}}$), the external field rotates so fast that the arc cannot synchronously follow any more and will start to asynchronously wobble without notable translation.

Because the precession of $\boldsymbol{e}_3$ plays a crucial role for the rotation-translation coupling, it is now clear that the orientation of the dipole moment with respect to the shape has dramatic consequences for the propulsion gait of the arc. The symmetry of the arc leaves two possibilities to affix a dipole to the body-coordinate-frame, which becomes apparent if both ends are colored differently (see Fig. 3.1a). The $\boldsymbol{e}_3$ axis either points towards the red or the green end, i.e. $\boldsymbol{e'}_3 \rightarrow -\boldsymbol{e}_3$ and $\boldsymbol{e'}_2 \rightarrow -\boldsymbol{e}_2$. However, the choice for solving the equations is purely arbitrary and color is obviously not a property that impacts the hydrodynamic behaviour. Consequently, if a precession of $\boldsymbol{e}_3$ around $\hat{z}$ leads to a propulsion for a specific choice of body coordinates, then a precession of $\boldsymbol{e'}_3$ yields another valid solution. A quantitative analysis of Eq. (3.5) for differently oriented magnetic dipole moments $\boldsymbol{m}$ reveals, that for the special case in which $\boldsymbol{m}$ is aligned in the plane that is spanned by $\boldsymbol{e}_1$ and $\boldsymbol{e}_3$ (see Fig. 3.1b), there are indeed two solutions to Eq. (3.5): $\boldsymbol{U} = \pm U_z$ [19]. They yield propulsion velocities in opposite directions $\pm\hat{z}$ and both have equal probability to materialize. Importantly, the dual solution, also called bistability, implies that the arc's propulsion direction is independent from the external fields rotation sense. For a clock-wise rotation sense of the external field it can translate with either the red or the green end in front. The dual solution becomes even more apparent by assuming an uncoloured arc with $\boldsymbol{m}$ *in-plane*, where one cannot distinguish in which direction the vector $\boldsymbol{m}$ points, see Fig. 3.1b. Two arcs, one where $\boldsymbol{m}$ points towards red and another where it points towards green are identical objects. On the contrary, in case the magnetic moment $\boldsymbol{m}$ has a component perpendicular to the $\boldsymbol{e}_1\boldsymbol{e}_3$-plane, as schematically depicted in Fig. 3.1c, then the two solutions collapse into one. Now the shape together with the vector $\boldsymbol{m}$ generates distinct objects, because it matters which way the dipole moment tilts, i.e. if it does have a component along $+\boldsymbol{e}_2$ or $-\boldsymbol{e}_2$. Such an arc will have a unidirectional propulsion gait, which means the external field's sense of rotation fixes the propulsion direction, similar to a corkscrew—a right-handed screw moves into $+z$, whereas the opposite enantiomer moves into reverse direction $(+z)$ upon a clock-wise rotation around $\hat{z}$.

Another important remark is that the torque that is exerted on the object could also be generated by an electric dipole moment $\boldsymbol{d}$ in an external electric field $\boldsymbol{E}$:

$$\boldsymbol{L} = \boldsymbol{d} \times \boldsymbol{E}. \tag{3.6}$$

This will not change the propulsion characteristics of the object as the hydrodynamic coupling is not sensitive to the nature of the field. Instead it will impact the transformation properties under parity and charge conjugation (see Sect. 2.5.1).

3.3 Macroscopic Arc-Shaped Propellers

The orientation of the dipole moment with respect to the body was predicted to play a crucial role for the propulsion characteristics. In contrast to micron sized V-shape propellers that are prone to imperfections during fabrication, a macroscopic model offers the possibility to precisely locate and unequivocally fix the orientation of the magnetic dipole moment $\boldsymbol{m}$ with respect to the body. Thus, macroscopic propellers have been examined initially, in order to prove the theoretical predictions. For this, an arc-shaped body was 3D-printed with an acrylic resin (*VeroClear*™) and it contained a small cavity into which a magnetic NdFeB (N45) cube with volume 1 mm^3 was glued as schematically depicted in Fig. 3.2a. Different orientations of $\boldsymbol{m}$ have been realized by printing several copies of the arc and re-aligning the cavity during the printing process as well as rotating the N45 cube properly. The body had a cross-sectional radius of $\ell = 1$ mm, a height of approximately $h = 17$ mm, a centerline radius of $r_c = 9.56$ mm and a central angle of 119°. As shown in the photograph in Fig. 3.2b, the arc was conveniently colored either completely red or half green and half red in order to distinguish the two ends by eye. To ensure low Re the arc was immersed in highly viscous glycerol with $\eta = 10^3$ cP, resulting in $Re \approx 1$. The high viscosity and small density mismatch between *VeroClear* ($\rho = 1.19\,\mathrm{g \cdot cm^{-3}}$) and glycerol ($\rho = 1.26\,\mathrm{g \cdot cm^{-3}}$) ensured that sedimentation played no role. The external field has been generated by two magnetic disks with an iron core, that provided a homogeneous and permanent magnetic field of 30 mT in between. They were mounted into a lathe that mechanically rotated the magnets around the **z**-axis, creating a rotating field in xy around a cuvette that was held between the magnets (see Fig. 3.2c). The cuvette was custom made by laser cutting and gluing together acrylic glass and had a size of $100 \times 95 \times 45$ mm^3 to fit into the empty space between the magnets. Typically, the arc was manually placed in the center of the cuvette before the external field rotation was switched on and moved until it approached the cuvette wall. Videos of the experiments have been acquired with a camera (*Canon* EOS 600D) that was mounted perpendicular to the lathe in the $x\,z$. Evaluation of

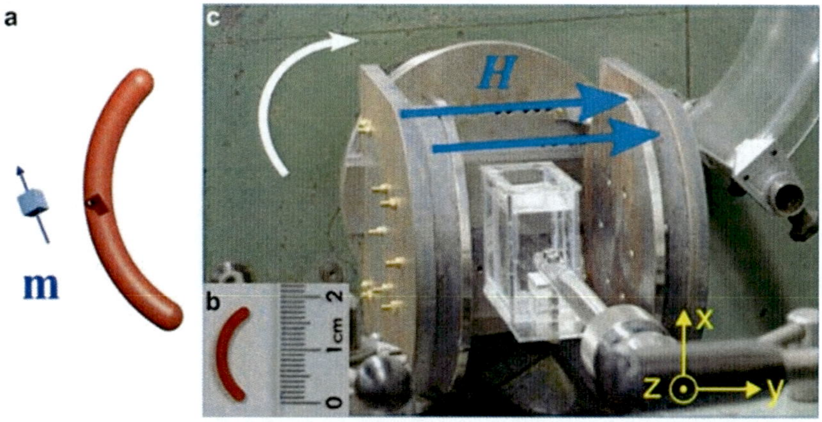

Fig. 3.2 Experimental setup for experiments with macroscopic models. **a** Schematic drawing of the 3D printed arc-segment that had a compartment to hold a small cubic (N45) magnet. **b** Photograph of the as-prepared cm-sized and magnetized arc, here colored red. **c** Lathe that holds two permanent magnets that are mechanically spun around a cuvette filled with highly viscous glycerol

the videos, i.e. tracking the arcs' position inside the cuvette over time, has been performed with python by first applying a color filter and then utilizing the openCV library to find the arc's contour and center within a single frame (details are described in Appendix A.3).

3.3.1 Achiral Objects with Arc-Shaped Body

First, achiral (parity-even) objects have been investigated. Parity ($\widehat{P}$) inverts the spatial coordinates of the shape (see Eq. (2.24)) as well as the position a magnetic dipole moment has (see Table 2.1). Importantly the direction of the magnetic dipole is unaltered after the execution of parity. Two different arcs with a magnetic moment aligned parallel to either the $\boldsymbol{e}_2$ or $\boldsymbol{e}_3$ axis have been fabricated. The transformation properties under parity of those two arcs can be deduced from the schematic drawings in Fig. 3.3. For both shapes the mirror image can be brought to coincide with the initial object and hence they are $\widehat{P}$-even (achiral). Both have been actuated by an external field rotating in the $x\,y$ and their z-coordinate was tracked over time with the results shown in Fig. 3.4. The data in Fig. 3.4a corresponds to the object shown in Fig. 3.3a that has a dipole moment along $\boldsymbol{e}_2$, and hence perpendicular to the "V"-body. It was thus rotating around the $\boldsymbol{e}_3$ axis and the external field had a rotation frequency of $f = 5.25$ Hz. The other object (Figs. 3.3b and 3.4b) had its dipole moment aligned along $\boldsymbol{e}_3$, and hence it spun around $\boldsymbol{e}_2$ exhibiting a higher hydrodynamic resistance. Consequently the step-out frequency is lower and it was actuated with $f = 1.05$ Hz only, because for higher frequencies it could not synchronously follow the external

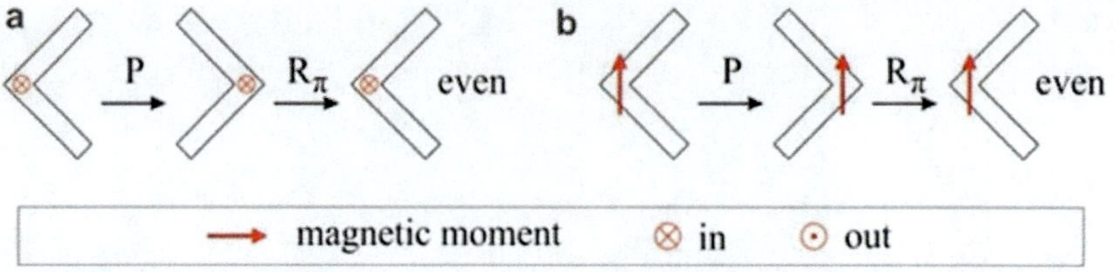

Fig. 3.3 V-shaped objects with their magnetic moment being aligned parallel to the principal axes of rotation, either along $\boldsymbol{e}_2$ (**a**) or $\boldsymbol{e}_3$ (**b**), as depicted in Fig. 3.1. Application of parity ($\widehat{P}$) and a subsequent rotation of 180° proves that both objects are $\widehat{P}$-even and thus achiral

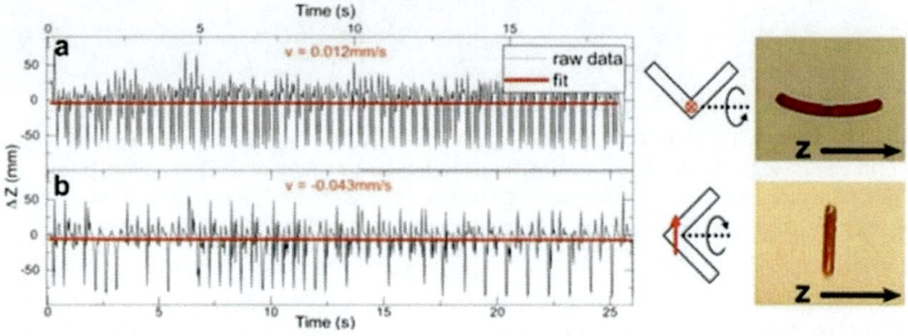

Fig. 3.4 Tracked z-Position of achiral V-shaped objects that have a magnetic moment aligned along a principal axis of rotation, reveals zero translational velocity. **a** Arc from Fig. 3.3a rotating with $f = 5.25$ Hz around $\boldsymbol{e}_3$ and **b** the arc depicted in Fig. 3.3b rotating with $f = 1.05$ Hz around $\boldsymbol{e}_2$. Adapted figure with permission from [1] Copyright 2018 by the American Physical Society

field any more. Both shapes did not precess around $\hat{z}$, as expected for a rotation along a principal axis of rotation. In Figs. 3.4 the comparably large magnitudes of the z-position in individual moments originate from erroneous position finding of the tracking algorithm and not because the arc changed its actual position in the cuvette (see Appendix A.3). By fitting a linear regression to the data the translational velocities were extracted to be $U_z = 0\ \text{mm} \cdot \text{s}^{-1}$ within the accuracy of the experiment. In agreement with the predictions from theory, the two presented achiral arcs are non-propulsive.

3.3.2 *Chiral Objects with Arc-Shaped Body*

Next, arc-shaped objects, which are chiral ($\widehat{P}$-odd) were examined. For this, the magnetic dipole moment was aligned along one of the arms, as visualized in Fig. 3.3b and a schematic drawing to deduce its symmetry under parity is shown in Fig. 3.5a. The latter proves clearly, that upon execution of parity one cannot find a proper rotation that superposes the object onto its initial state, and hence it is chiral. Note, that the dipole moment is not parallel to one of the principal axes of rotation, however

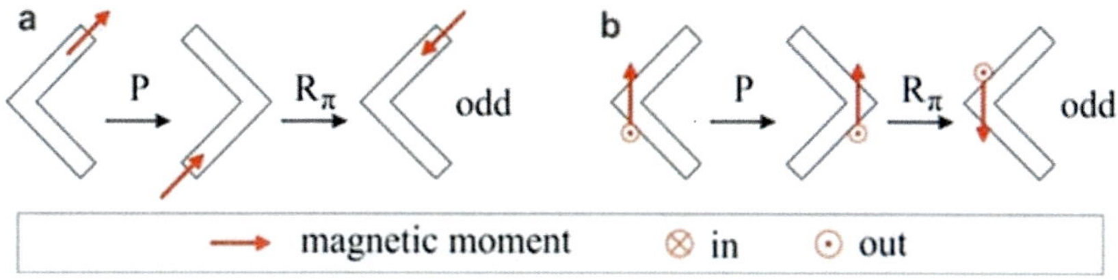

Fig. 3.5 V-shaped objects with an arbitrarily aligned magnetic moment that is not parallel to the principal axes of rotation. **a** Moment parallel to $\boldsymbol{e}_1\boldsymbol{e}_3$ and pointing along one of the arms of the V-shape. **b** The dipole moment is tilted out of the symmetry plane and has a component along $\boldsymbol{e}_2$. Application of parity ($\widehat{P}$) and a subsequent rotation of 180° reveals that both objects are $\widehat{P}$-odd and therefore chiral

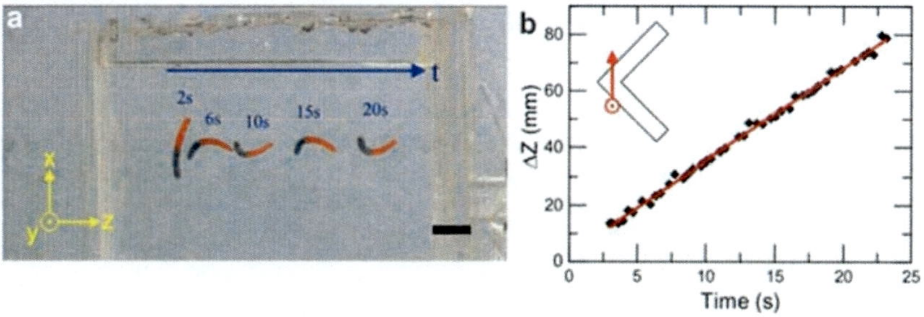

Fig. 3.6 Propulsion gait of a macroscopic chiral arc **a** Image overlay of several frames showing that after a short time the arc undergoes a stable precession around the $\hat{z}$ axis resulting in a constant translation along z (scale bar 1 cm). **b** Plotting the displacement Δz versus time indicates that the arc exhibits a constant velocity of $U_z = 3.3\ \mathrm{mm \cdot s^{-1}}$. Adapted figure with permission from [1] Copyright 2018 by the American Physical Society

it lies in the $\boldsymbol{e}_1\boldsymbol{e}_3$-plane. Another arc was realized that had a dipole moment tilted out-of-plane, i.e. $\boldsymbol{m}$ had a component along $\boldsymbol{e}_2$. This corresponds to the object shown in Fig. 3.3c and the application of parity on this object is illustrated in Fig. 3.5b. Again, the object is $\widehat{P}$-odd and therefore chiral.

The object from the latter case was actuated with a rotating field ($f = 1.5$ Hz) and Fig. 3.6a shows an image overlay of several frames taken from the captured video. It was observed that in the beginning, when the external field has been static, the arc was stationary and $\boldsymbol{m}$ aligned with the external field lines. After the rotational motion of $\boldsymbol{H}$ in xy was switched on, the object tumbles for the first few seconds before it starts a stable precession around $\hat{z}$. Simultaneously it starts to translate along $\hat{z}$. Evaluation of the objects' position over time reveals a constant propulsion velocity of $U_z = 3.3\ \mathrm{mm \cdot s^{-1}}$, as shown in Fig. 3.6b. Upon a reversal of the external field rotation sense, the arc was moving in the opposite direction. The same experiment was repeated with the second object (Fig. 3.5a) that had a dipole moment pointing to the tip of one arm, and again a stable propulsion along $\hat{z}$ has been observed. This behaviour is predicted by theory and shows that, a highly symmetrical body like the arc can indeed be propelled by means of rotation-translation coupling at low Re if

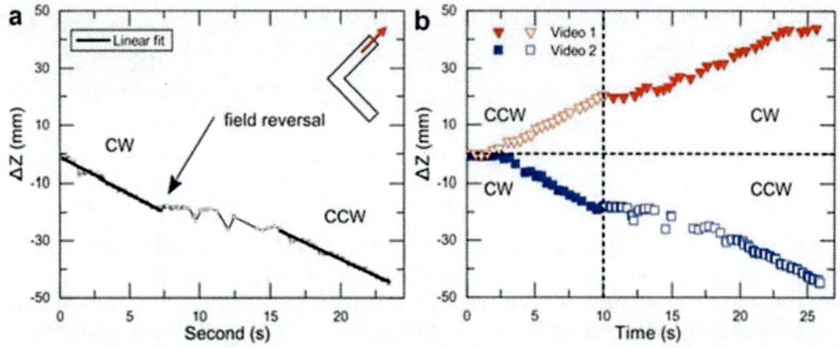

Fig. 3.7 Displacement of the magnetized object that has a magnetic moment that lies in the $\boldsymbol{e}_1\boldsymbol{e}_3$-plane. **a** After the external fields' rotation sense is reversed, the object tumbles a few seconds before it translates again in the same direction ($-\hat{z}$) with the same velocity as before. **b** Two videos acquired for the same object with either clockwise (CW) or counter-clockwise (CCW) rotation sense followed by a reversal of the rotation sense after 10 s (vertical dashed line). The displacement in both experiments shows that one and the same object can travel in $\pm\hat{z}$ and maintain its propulsion velocity and direction, regardless of the external fields' rotation sense. Reprinted figure **b** with permission from [1] Copyright 2018 by the American Physical Society

the magnetic moment does not point along one of the principal axes of rotation and consequently the body is spun in a non-trivial way.

However, the theory yields a dual solution and hence bistable propulsion characteristic for the latter case. As mentioned above, this implies that the arc's direction of translation is independent of the external field's rotation sense. To verify this hypothesis the same object was rotated by an external field ($f = 1.25$ Hz) with a clockwise rotation sense and then the rotation sense was abruptly reversed (to counter-clockwise). The resulting velocity diagram is shown in Fig. 3.7a and it is seen that after the reversal the arc tumbles a few seconds, before it translates with the same velocity as before along the identical direction. Note, that a reversal of the external field's rotation sense mathematically reverses the angular velocity vector: $\boldsymbol{\Omega} = \omega\hat{z} \rightarrow -\boldsymbol{\Omega} = -\omega\hat{z}$. Consequently, the arc was first travelling along $-\hat{z}$ and after the rotation reversal along $+\hat{z}$. Further, it was validated that the same object can first travel along $+\hat{z}$ and then continue to move in $-\hat{z}$ direction for a field that rotates counter-clockwise and subsequently clockwise (see Fig. 3.7b). This clearly proves that this object exhibits a bidirectional motion along the rotation vector and hence it can move in $\pm\hat{z}$ no matter what the external fields rotation direction is.

To deduce the dependency between propulsion velocity U_z and the external fields' rotation frequency ($f = 2\pi\omega^{-1}$), several measurements with different actuation frequencies were performed. Both chiral objects depicted in Fig. 3.5 were examined for clockwise and counter-clockwise actuation and the expected uni- and bidirectional propulsion gaits for the different orientations of $\boldsymbol{m}$ are observed (Fig. 3.8). The unidirectional behaviour for the tilted alignment of $\boldsymbol{m}$ yields a propulsion velocity that is always in the direction to the angular velocity vector $\boldsymbol{\Omega}$ and thus always positive. In contrast, the alignment of $\boldsymbol{m}$ in the $\boldsymbol{e}_1\boldsymbol{e}_3$-plane exhibits arbitrary motion direction and

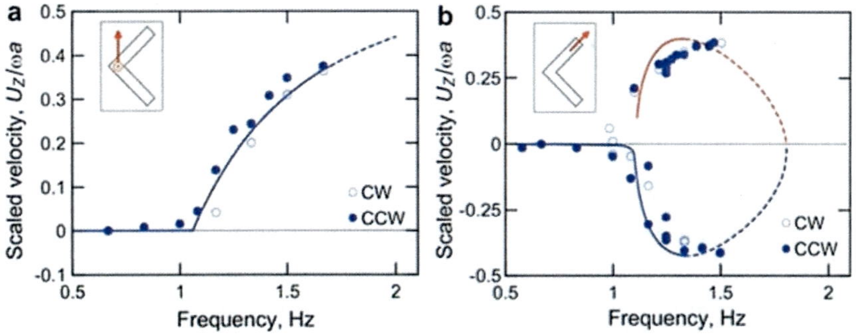

Fig. 3.8 Scaled propulsion velocity U_z versus the external field's rotation frequency $f = 2\pi\omega^{-1}$ for chiral objects. Open and closed symbols, respectively, correspond to a clockwise and counter-clockwise rotation sense of the external magnetic field, whereas solid lines are calculated from theory. **a** The object that has its moment tilted out-of the $\boldsymbol{e}_1\boldsymbol{e}_3$-plane exhibits unidirectional propulsion, similar to a corkscrew. **b** If $\boldsymbol{m}$ lie in the $\boldsymbol{e}_1\boldsymbol{e}_3$-plane then the direction of U_z is not fixed by the external fields' rotation sense and hence bistable propulsion occurs. Adapted figure with permission from [1] Copyright 2018 by the American Physical Society

hence, positive and negative velocities occur irrespective of the external field's rotation sense. The individual measurements shown in Fig. 3.8 were conducted over one day from morning to evening (in a busy machine workshop). This led to an increase in temperature and humidity of the environment. Because the glycerol viscosity is sensitive to those parameters, the measurements acquired at later times were scaled by a constant factor of 1.4, in order to account for this change of environmental parameters.[3] The theoretical dependency between U_z and actuation frequency f was calculated by solving Eq. (3.2) for the actual parameters of the experiment. Geometric dimensions of the arc are described above (see Sect. 3.3), whereas the mobility tensors $\mathscr{F}$ and $\mathscr{G}$ were obtained numerically by A. Leshansky and his colleagues [1]:

$$\mathscr{F} = \frac{10^{-4}}{\eta\ell^3}\begin{pmatrix} 2.36 & 0 & 0 \\ 0 & 2.42 & 0 \\ 0 & 0 & 14.95 \end{pmatrix}, \quad \text{and} \quad \mathscr{G}_{23} = \frac{2.14 \cdot 10^{-4}}{\eta\ell^2}. \tag{3.7}$$

Further, the tumbling-to-wobbling as well as the step-out frequency were estimated from the measured data (Fig. 3.8). By substituting all these numbers into Eq.(3.2) the functional dependency between U_z and f is theoretically described as [1]

$$\frac{U_z}{\omega\ell} = 0.614\left(1 - \frac{1.06^2}{f^2}\right), \tag{3.8}$$

for the unidirectionally propulsive object in Fig. 3.8a that had a magnetic moment with a non-zero component along $\boldsymbol{e}_2$. Conversely, the bidirectional propulsion char-

[3] For the 99.5% pure glycerol that has been used, a temperature increase from 25°C to 29°C corresponds to a decrease in dynamic viscosity η by a factor 1.4.

acteristics of the arc which only had a magnetization in the $\boldsymbol{e}_1\boldsymbol{e}_3$-plane (Fig. 3.8b) yields:

$$\frac{U_z}{\omega\ell} = \pm 0.775\sqrt{1 - \frac{1.10^2}{f^2}} \cdot \sqrt{\frac{1.80^2}{f^2} - 1}. \tag{3.9}$$

Both functions are plotted as solid lines in Fig. 3.8a and b, respectively, and they show excellent agreement with the measured data. Consequently, the uni- and bistable propulsion characteristics for chiral objects agree quantitatively as well as qualitatively with the theoretically predicted behaviour.

3.4 Symmetry Analysis of Magnetized Propellers

The results shown so far verify that geometrically achiral objects, i.e. objects that have an achiral shape, can still be propulsive by exploiting rotation-translation coupling at low Re. However, it was demonstrated that, if bodies with those shapes are propulsive, they are a chiral objects when the magnetic dipole moment is taken into consideration (as it should) for the symmetry analysis (see Fig. 3.5). On the contrary, in cases for which the object was achiral, the dipole moment was parallel to one of the principal axes of rotation ($\boldsymbol{m} \parallel \boldsymbol{e}_i$), and consequently the object is non-propulsive (see Fig. 3.3). In fact, it can be mathematically proven that any $\widehat{P}$-even magnetized object either has $\boldsymbol{m} \parallel \boldsymbol{e}_i$ or that its rotation-translation coupling tensor vanishes $\boldsymbol{C} = \mathscr{G}^{-1} = 0$, regardless of its shape [1].

This can be understood if one realises that by the definition of parity (Eq. (2.24)), $\widehat{P}$-even objects are mirror-symmetric and thus a rotation can be found that superposes the object to the initial state after parity was executed. In other words, a proper rotation exists for $\widehat{P}$-even objects, whose action is equivalent to the execution of parity: $\widehat{R} \equiv \widehat{P}$. Because the rotation mobility and coupling tensors, respectively, are even and odd under parity,

$$\widehat{P}\mathscr{F} = \mathscr{F} \quad \text{and} \quad \widehat{P}\mathscr{G} = -\mathscr{G}, \tag{3.10}$$

it follows for an arbitrary magnetized object, that [1]:

$$\begin{aligned} \widehat{R}\mathscr{F}\widehat{R}^{-1} &= \mathscr{F} & (i), \\ \widehat{R}\mathscr{G}\widehat{R}^{-1} &= -\mathscr{G} & (ii), \\ \widehat{R}\boldsymbol{m} &= \boldsymbol{m} & (iii). \end{aligned}$$

Because of (i), $\widehat{R}$ is diagonal in the body-coordinate-frame of principal axes $\boldsymbol{e}_1, \boldsymbol{e}_2, \boldsymbol{e}_3$. If $\widehat{R} = \mathbb{1}$ then from relation (ii) it follows that $\mathscr{G} = 0$, and hence $U_z = 0$. In any other case, $\widehat{R}$ must be a rotation by π around one of the principal axes $\boldsymbol{e}_i$ and then, due to relation (iii), $\boldsymbol{m}$ has to be along this rotation axis $\boldsymbol{m} \parallel \boldsymbol{e}_i$, and again $U_z = 0$. In

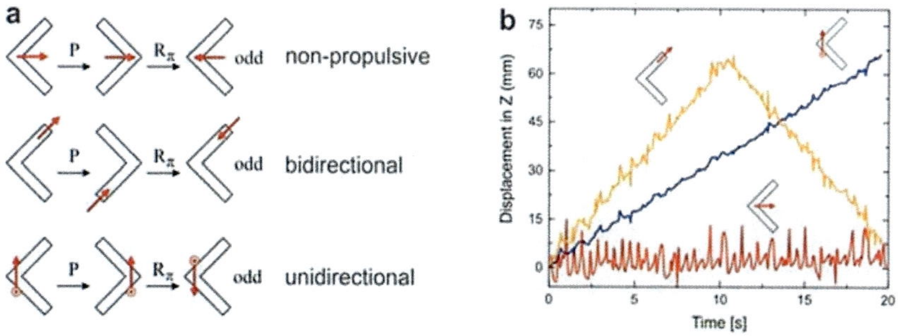

Fig. 3.9 Comparison of V-shaped objects whose magnetic dipole moment $\boldsymbol{m}$ is aligned parallel to the principal axes of rotation $\boldsymbol{e}_1$ or whose moment lies in the $\boldsymbol{e}_1\boldsymbol{e}_3$-plane or is tilted out of this plane. **a** Schematic drawings illustrate that all objects are chiral ($\widehat{P}$-odd) but exhibit either non-, bi-, or unidirectional propulsion characteristics. **b** Schematically depicted experimental velocity diagrams providing the displacement over time for the shapes corresponding to **a**

consequence, $\widehat{P}$-even (achiral) objects that are actuated by an external magnetic field can never be propulsive. Remarkably, this proves that the magnetic 3-bead cluster [17] mentioned in the beginning (Sect. 3.1) must be a chiral object, at least if it is propulsive.

Interestingly, one can imagine a magnetized shape that is chiral and non-propulsive. If $\boldsymbol{m}$ is parallel to $\boldsymbol{e}_1$ (according to Fig. 3.1) it will rotate around a principal axes of rotation, and the schematic drawing, which is shown in the top panel of Fig. 3.9a, reveals that this object is $\widehat{P}$-odd. For comparison the two other shapes that represent the unidirectional and bidirectional propulsion gait are drawn next to it. In Fig. 3.9b the expected displacement in z is shown, respectively. This example clearly indicates that chiral objects can be non-propulsive too, which demonstrates that chirality is not a pre-requisite for propulsion by rotation-translation coupling.

However, if chirality alone does not determine propulsion characteristics of objects that exploit rotation-translation coupling, then the question remains if there is a feature that can do so? With the help of the presented experiments a rigorous symmetry scheme has been developed to answer this question [1]. It reveals that not only parity $\widehat{P}$ but also charge conjugation $\widehat{C}$ has to be considered. Then, it is possible to predict whether a shape obeys uni-, or bidirectional propulsion or if it is not propulsive at all.

The symmetries of a magnetic-dipolar body with two mutually perpendicular symmetry planes have been examined. Schematically this is depicted for a V-shaped body in Fig. 3.10. Panel a) shows that the V-shape, which has its moment aligned perpendicular to the $\boldsymbol{e}_1\boldsymbol{e}_3$ is achiral ($\widehat{P}$-even) because a rotation around $\boldsymbol{e}_2$ (denoted $\widehat{R}_2$[4]) would return the object to its initial state. By first executing parity and subsequently charge conjugation $\widehat{C}$, the direction of the dipole moment would be reversed as described in Sect. 2.5.1 (Table 2.1), but the object can still be rotated back to its initial state, e.g. by $\widehat{R}_2\widehat{R}_1$. Thus the object is not only $\widehat{P}$- but also $\widehat{C}\widehat{P}$-even. The

[4] The index i in $\widehat{R}_i$ indicates a rotation about π around the i-th principal axis of rotation $\boldsymbol{e}_i$.

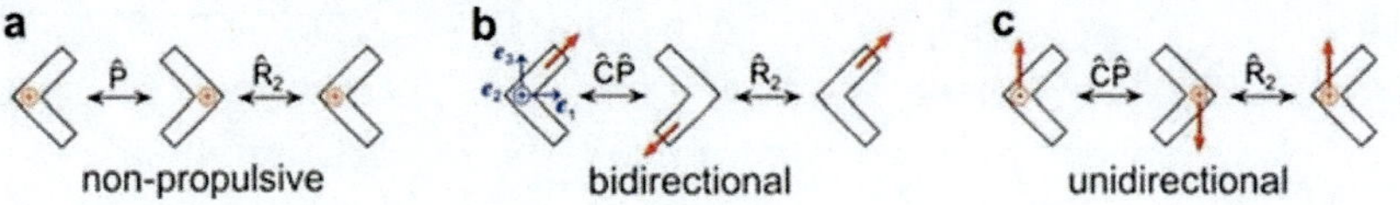

Fig. 3.10 Symmetries of an object with two mutually perpendicular symmetry planes and that is carrying a magnetic dipole moment (red) under parity ($\widehat{P}$) and charge conjugation ($\widehat{C}$). **a** $\widehat{C}\widehat{P}$- and $\widehat{P}$-even object. **b** $\widehat{C}\widehat{P}$-even and $\widehat{P}$-odd object (principal axes of rotation in blue). c) $\widehat{C}\widehat{P}$- and $\widehat{P}$-odd object. Adapted figure with permission from [1] Copyright 2018 by the American Physical Society

$\boldsymbol{e}_1\boldsymbol{e}_3$-magnetized body (Fig. 3.10b) is found to be $\widehat{C}\widehat{P}$-even, but it is $\widehat{P}$-odd upon application of parity alone (see Fig. 3.5a). The third object is the V-shape for which $\boldsymbol{m}$ is titled out of the symmetry plane, and it is shown that this objects is $\widehat{C}\widehat{P}$-odd as well as $\widehat{P}$-odd (Figs. 3.10c and 3.5b). The latter case is thus the least symmetric and it can be mathematically proven that for this case unidirectional propulsion occurs [1]. Higher symmetry ($\widehat{C}\widehat{P}$-even and $\widehat{P}$-odd) yields bidirectional propulsion characteristics, whereas above it was already proven that highly symmetrical $\widehat{P}$-even magnetized objects are always non-propulsive. Thus the prediction whether or not an object is propulsive and if it is uni- or bidirectional can be made by utilizing CP-symmetry in conjunction with parity.

It is important to mention that here the symmetries were derived in the context of a magnetic dipole in a magnetic field. However, in Sect. 2.5.1 it was explained that electric fields exhibit the exact opposite transformation properties under parity and the same symmetry under charge conjugation (Table 2.1). In case an object would be driven by an external electric field which exerts a torque on a permanent electric dipole, the same dynamics as described would apply, but the symmetries are swapped. In fact, one needs to rewrite every operator in Fig. 3.10 with its counterpart: $\widehat{C}\widehat{P} \iff \widehat{P}$. Consequently, the objects in Fig. 3.10b and c would be $\widehat{P}$-even and therefore truly achiral. Changing the nature of the actuating field will not change the propulsion characteristics, which rely on (hydro-)dynamics rather than just object symmetries. Thus a truly achiral object can potentially be propulsive due to rotation-translation coupling, if it is actuated by an electric field.

3.5 Microscopy of Nanoscale V-Shaped Propellers

The previous experiments with macroscopic propellers verified that a chiral object can be either non-, bi- or unidirectional propulsive. Additionally, the question if a truly achiral object can be made propulsive by means of rotation-translation coupling, has not been addressed. With magnetically driven objects this is impossible, because achiral magnetized objects will always have $U_z = 0$, as mathematically proven in Sect. 3.4. On the other hand, theory suggested that for electrically driven objects it is possible to do so. Experiments utilizing macroscopic objects are easier implemented with magnetic structures, as it is harder to obtain a macroscopic electric dipole that is not screened. Although some ferroelectric materials exist, such as $BaTiO_3$, $LiNbO_3$, or $Pb(Zr_xTi_{1-x}O_3)$, their ferroelectricity behaviour is often limited to thin-films (due to poling) or micrometer sized crystals. Moreover the attainable dipole moments are rather small, and consequently the exerted torques are also small. A simple experiment was tried with a $1\,mm^3$ cube of periodically poled $LiNbO_3$, which was exposed to an external electric field with strengths up to $100\,kV \cdot m^{-1}$. The external field did not exert a large enough torque onto the macroscopic $LiNbO_3$ cube.

However, the validity of the symmetry arguments could be verified for microstructures which are electrically polarizable and fabricated via GLAD (Sect. 2.7). First, magnetic V-shape microstructures were used to confirm the findings of the previous section. Thereafter it is demonstrated that it is indeed possible to exert a torque on microstructures, due to an induced electric dipole originating from polarization effects. To observe the microstructures' motion an optical microscope (*Zeiss* Axio Observer) was used. The magnetic field was generated using three Helmholtz coil pairs that are orthogonally arranged to each other, with the sample placed in its center. Each pair of coils generated a homogeneous magnetic field along one dimension and the superposition of magnetic fields along each of the three spatial directions allowed an arbitrarily oriented magnetic field vector to be generated. The coil setup was in-house engineered and controlled by a custom written python software. The setup is described in greater detail in Appendix A.2. The electric field has been generated with a simpler setup in two dimensions only. It consisted of 4 metal electrodes in the sample plane that were driven by two synchronized function generators, which will be described below. In all of the experiments the grown structures are around $\simeq 3\,\mu m$ in size and were actuated in water ($\eta \simeq 1$ cP), yielding a Reynolds number of $Re \lesssim 10^{-11}$. Videos of the microscope experiments have been acquired with a CCD camera (*Andor* iXon) and were subsequently evaluated with python utilizing the Trackpy package or ImageJ.

3.5.1 Magnetically Driven Chiral V-Shape

Magnetized V-shaped microsctructures were fabricated by GLAD. A thin layer of Ni has been grown on a monolayer of 500 nm diameter SiO_2-beads, that acted as a nucleation seed layer. On top of the Ni, a SiO_2 column was grown perpendicular to the Si-substrate, utilizing fast substrate rotation. This was followed by a second SiO_2 column grown without rotation of the substrate, tilted with respect to the first to form the V-shape. An SEM image of the resulting structure is shown in Fig. 3.11a, where the magnetic Ni section is seen (lighter color). After fabrication, the V-shapes were magnetized on the substrate by placing them in a strong magnetic field of $\boldsymbol{H} = 1.8$ T. Because the orientation of the V-shape on the substrate is known it was possible to induce a specific alignment of $\boldsymbol{m}$ with respect to the body by properly orienting the substrate during magnetization. In doing so the V-shapes have been magnetized with $\boldsymbol{m}$ either parallel to the $\boldsymbol{e}_1\boldsymbol{e}_3$ or orthogonal to it (see Fig. 3.11b). This

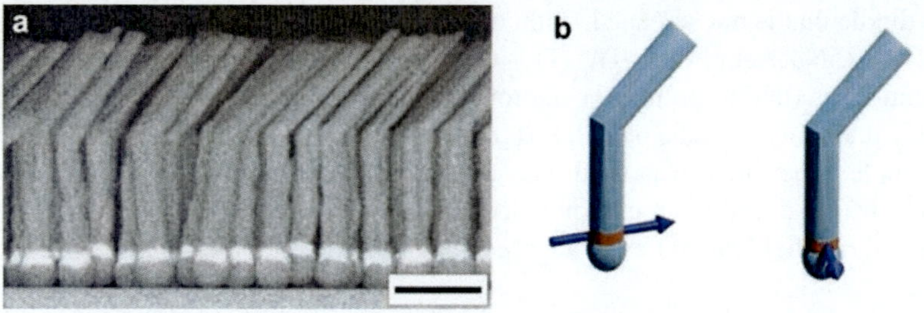

Fig. 3.11 V-shaped microstructures fabricated with GLAD. **a** SEM image shows the SiO_2 columns that form a "V" with magnetic section (white) consisting of Ni (scale bar 1 μm). **b** Schematics for $\boldsymbol{m}$ being aligned parallel or orthogonal to the $\boldsymbol{e}_1\boldsymbol{e}_3$-plane, which is in practice controlled by magnetizing the structures in a strong magnetic field on the substrate. The left object is $\widehat{C}\widehat{P}$-even and $\widehat{P}$-odd, whereas the right is $\widehat{C}\widehat{P}$- and $\widehat{P}$-even

renders the micro V-shape chiral ($\widehat{C}\widehat{P}$-even and $\widehat{P}$-odd) or achiral $\widehat{C}\widehat{P}$-even and $\widehat{P}$-even, respectively. After magnetization the structures were immersed in a solution of 150 μM poly(vinylpyrrolidone) via ultrasonication. Both types (achiral and chiral) were observed in the microscope under a rotating external field ($B = 60$ G) with frequency $f = 25$ Hz. The position of several microstructures were tracked and the displacement along the angular velocity vector $\hat{z}$ over time is shown in Fig. 3.12a for the in-plane magnetized and Fig. 3.12b for the orthogonally magnetized shape. As expected, the former exhibits velocities ($U_z \approx 2.7\,\mu\text{m}\cdot\text{s}^{-1}$) with opposite sign, i.e. a bidirectional motion. Slight variations of the velocities are anticipated because of imperfections during the fabrication process and small deviations of the magnetic moment alignment. The latter yields $U_z \approx 0\,\mu\text{m}\cdot\text{s}^{-1}$ and is thus non-propulsive due to a magnetic moment which is aligned along a principal axis of rotation ($\boldsymbol{e}_2$). Both results are in agreement with the findings of the corresponding macroscopic V-shapes and are explained by the symmetry analysis in Sect. 3.4.

Additionally, it was demonstrated that the in-plane magnetized object (Fig. 3.12a) can be steered in 3D by turning the external field's plane of rotation between the three principal planes. Most noteworthy, a propulsion against gravity, out of focus of the microscope was achieved. This is shown in Fig. 3.12c, which visualizes a time-lapse sequence of microscope images together with a schematic drawing of the particles' position in z.[5] First, the object was in the microscope's focal plane and after switching on the external field it moved upwards, against gravity. After 19 s the field was switched off and the object slowly sedimented back into the focal plane. The on-off sequence was repeated one more time (not shown here). This experiment qualitatively proves that the micro-scale V-shape successfully converts a rotation into linear motion, away from a surface and is in agreement with the expectations.

[5] Here z indicates the optical axis of the microscope.

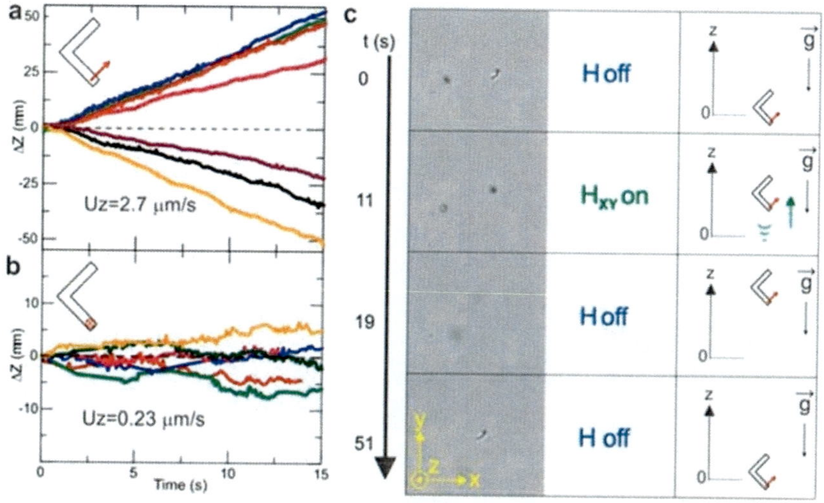

Fig. 3.12 Propulsion characteristics of V-shaped microstructures with different magnetization, i.e. alignments of $\boldsymbol{m}$ with respect tot the body. **a** Displacement along the angular velocity vector $\hat{z}$ of several chiral microstructures show a constant translation with average velocity $U_z \approx 2.7\,\mu\text{m}\cdot\text{s}^{-1}$ into opposite directions. **b** Achiral V-shapes show negligible displacements and are thus non-propulsive ($U_z \approx 0\,\mu\text{m}\cdot\text{s}^{-1}$). **c** The same object from **a** can also be steered against gravity out of focus of the microscope upon an externally rotating field in xy, and sediments back when the field is switched off after 19 s. Adapted figure with permission from [1] Copyright 2018 by the American Physical Society

3.5.2 *Electrically Driven Achiral V-Shape*

An achiral and propulsive microswimmer (by means of rotation-translation coupling) has never been reported, although previous reports mistakenly claimed this [17]. In Sect. 3.4 it is described that only an electrically actuated V-shaped structure can be (truly) achiral. To test the hypothesis that such an object can be propulsive too, another V-shape was fabricated by GLAD. This time, not with a magnetic Ni section. Instead one arm of the V-shape was covered with Au, which is highly polarizable in an electric field $\boldsymbol{E}$. This leads to an induced electric dipole moment $\boldsymbol{p} = \boldsymbol{\alpha} \cdot \boldsymbol{E}$ along one of the arms of the V-shape as schematically depicted in Fig. 3.13a. The symmetry of this object is $\widehat{P}$-even ($\widehat{C}\widehat{P}$-odd) and thus achiral. First, a SiO_2 column was grown perpendicular to the substrate, again with SiO_2 beads as a seed layer and fast substrate rotation. It was then coated on one side by Au utilizing a steep incident angle. Afterwards the second arm of the "V" was formed by a SiO_2 column, evaporated without substrate rotation. A SEM image of the resulting structure is depicted in Fig. 3.13b, which verifies that one of the arms is gold coated.

In order to generate a rotating electric field a four-electrode setup was built. The electrodes were made of 4 metal wires with diameter 100 μm that are glued into a microscope sample chamber, which consisted of two coverslips sealed with vacuum

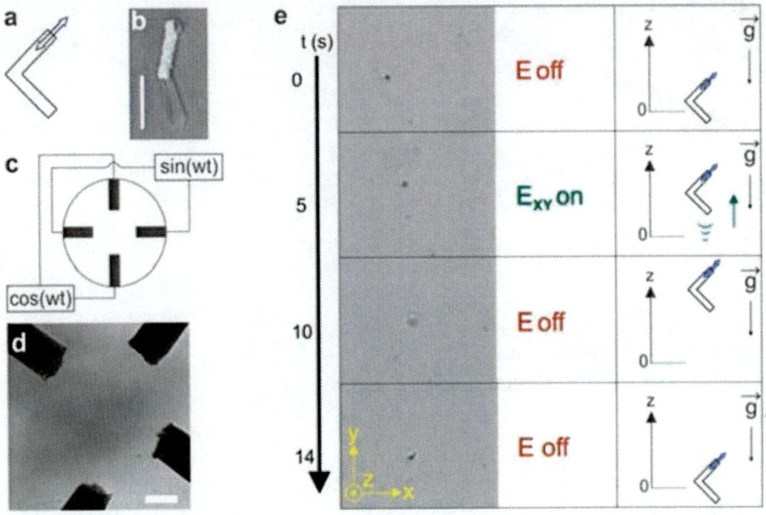

Fig. 3.13 Propulsion of an electrically actuated V-shape microstructure. **a** Schematic drawing of a V-shape that is polarizable along one arm reveals that this object constitutes a $\widehat{P}$-even and therefore (truly) achiral object. **b** SEM image of the microsctructure which consists of SiO_2 and has one arm coated with Au (white), fabricated with GLAD (scale bar 1 μm). **c** Schematics of two function generators, which are 90° phase-shifted and that drive a four-electrode setup in order to create a rotating electric field in the $x\,y$. **d** Microscope image of the four electrodes show a small misalignment possibly resulting in a gradient in xy (scale bar 100 μm). **e** The rotating electric field propels the V-shape against gravity, out of focus of the microscope and sediments back into the focal plane after the field was switched off after 10 s (images are artificially centered to adjust for the motion in xy). Adapted figure with permission from [1] Copyright 2018 by the American Physical Society

grease. Two function generators that were phase-shifted by 90° applied a sinusoidal potential at two orthogonal pairs of electrodes. A schematic illustration is depicted in Fig. 3.13c and a microscope image of the electrodes in Fig. 3.13d. As seen, the electrodes were slightly misaligned leading to a small gradient in xy. The applied electric field strength was $\boldsymbol{E} \approx 7.5\,\mathrm{kV \cdot m^{-1}}$ with an AC frequency of 500 kHz, leading to a rotating electric field in xy and along z. If the polarizable V-shape is placed in this field, a dipole is induced along the gold-coated arm. The polarizability $\boldsymbol{\alpha}$ is generally a tensorial quantity, and the shape anisotropy of the Au results in an anisotropic polarizability which is stronger parallel to the longer dimension, i.e. parallel to the arm ($\alpha_{\parallel} > \alpha_{\perp}$). Thus, an external electric field will exert a torque onto the gold-coated arm that will align it with the direction of $\boldsymbol{E}$ in the $x\,y$:

$$\boldsymbol{L} = \boldsymbol{p} \times \boldsymbol{E} = (\boldsymbol{\alpha} \cdot \boldsymbol{E}) \times \boldsymbol{E} = (\alpha_{\parallel} - \alpha_{\perp}) \cdot (\boldsymbol{n} \times \boldsymbol{E})(\boldsymbol{n} \cdot \boldsymbol{E}). \tag{3.11}$$

The external field's rotation frequency of 500 kHz led to a rotation of the object with a few Hz, in agreement with other studies observing electro-rotation of shape-anisotropic polarizable objects [21]. The observed propulsion of the electro-rotated object is shown in Fig. 3.13e. Note the two different colours in the microscope images, due to the different refractive indices of the arms. Similar to the magnetic counterpart (Fig. 3.12c) the particle was initially resting in the focal plane of the microscope. After the external rotating field was switched on, it moved along z, out of focus until the field was stopped. It then sedimented back into focus due to gravity and again, this sequence was repeated (not shown here). This experiment unequivocally demonstrated the ability of this object to translate upon a rotation, in accordance with the previous results. The similarity of the experiments between the electrically driven V-shape and it's magnetic counterpart is substantial, but importantly here the object is $\widehat{P}$-even and thus achiral. As mentioned in the symmetry analysis of Sect. 3.4 a magnetized $\widehat{P}$-even object will never be propulsive. Consequently, the reported electrically actuated V-shape is the first realization of a truly achiral object that is propulsive by means of rotation-translation coupling. This experiment provides a clear example that the motion of highly symmetrical objects at low Re depends on the dynamics rather than just symmetry.

3.6 High Density High Symmetry Swimmers at Low Re

In the previous sections it has been shown that even achiral ($\widehat{P}$-even) objects can be propulsive by means of rotation-translation coupling. It is therefore of interest to investigate shapes with even higher symmetries. A V-shape possesses two mutually perpendicular symmetry planes and thus the corresponding rotation-translation coupling tensor still exhibits two nonzero elements which are equal (see Eqs. (2.14) and (3.5)). However, for objects exhibiting even higher symmetry, e.g. a spherically isotropic body, it can be shown that the coupling tensor vanishes, i.e. $C^{-1} = \mathcal{G} = 0$ [20]. It follows that a sphere cannot be propelled by a externally-induced rotation-translation coupling mechanism at low Re. However, chemically self-powered microswimmers, i.e. catalytically-active Janus particles, can exhibit a self-propulsive motion owing to different chemical reactivities across their surface. The symmetry of the Stokes equation (Eq. (2.5)) is broken due to the distinct properties inherent to the different materials. One advantage of chemically powered microswimmers is that they can be very small and propel autonomously because they only need a chemical substrate ($f$uel) that powers their active motion, rather than an external stimulus. Obviously, a spherical object that does not carry a dipole moment is always $\widehat{P}$-even and therefore achiral.

Many different studies have demonstrated the self-propulsion of *chemically powered* micro- and nanoswimmers, comprising a large variety of different material systems and chemical reactions [22–28]. Physically powered swimmers examined during the previous sections rely on a external torque that is exerted and that scales with the amount of magnetic or polarizable material, and hence the ability to actuate the microswimmers depends on their volume, i.e. their geometric dimensions. Micro- and nanoswimmers at very small sizes are interesting, because only sub-micron particles are able to penetrate biological tissue [29]. The smallest physically-powered swimmer reported to date that moves *via* rotation-translation coupling is still about $\ell \sim 400$ nm in length [12]. Also, if the size is further reduced, the fabrication of complex 3D-shapes becomes more challenging and the impact of Brownian forces onto particles with a few tens of nanometers in size will be very strong. Consequently this would perturb the stable and directed propulsion exhibited by physically powered nanoswimmers [11]. Conversely, chemically powered swimmers have been demonstrated to be still propulsive with sizes down to $\ell \sim 30$ nm [27] and fabrication of nanoscale bodies exhibiting spherical symmetry is in general much simpler. On the downside, self-propelling particles usually exhibit mainly Brownian motion with only some drift rather than a directed motion, but guiding the particles' with the help of an external magnetic field was demonstrated as well [28].

Although the research topic of (highly symmetric) self-propelled particles has attracted a lot of attention in the recent years, the details of the underlying chemical reactions are often unresolved yet. Nearly all experiments to date examined micrometer-sized particles and observed their velocity under a microscope, e.g. for varying fuel concentrations. Typically their MSD was evaluated and later described theoretically by a simple model [30]. However, because of the comparably large particle sizes the Janus particles rapidly sediment to the bottom of the microscope sample chamber and therefore such experiments were always performed in close proximity to a surface. The surface properties like the zeta-potential, the contact angle, etc. will generally influence the chemical environment and just recently this has been demonstrated to impact the self-propulsion and therefore the velocity a Janus swimmer exhibits [31]. Consequently, the exact propulsion mechanism remains ambiguous, because different chemical potentials would lead to different reactions. However, the details are still subject of an ongoing debate, since numerical simulations and theoretical predictions are contradictory [31–35], while experimental studies provide inconsistent results [35–38]. The latter is very likely exacerbated by highly non-uniform experimental boundary conditions, e.g. the wall-effects have been overlooked or preparation protocols differ considerably. Especially for Janus particles with small sizes ($\ell \lesssim 500$ nm) it is an open question whether or not bulk reactions, i.e. the dissociation and recombination of water molecules ($2H_2O \rightleftharpoons H_3O^+ + OH^-$), play a role for their propulsion [35]. It was thus an interesting task to examine such shapes and acquire experimental data for JPs with radii between $r \sim 50 - 500$ nm. Another interesting feature is that particles in this size range would bridge the gap between naturally occurring swimmers with sizes up to $r \lesssim 30$ nm (e.g. enzymes) and nearly all artificial self-propelling particles reported to date, which normally have sizes of $r > 500$ nm.

One possible reason why Janus particles with small sizes have thus far not been examined is that their observation is hardly possible by conventional bright-field microscopy due to the diffraction limit.[6] In order to do so, scattering based techniques have to be employed. One idea would be to use dark-field microscopy and standard particle tracking to perform the usual MSD evaluation methods. But dark-field microscopy has a limited depth of focus and small objects move very fast in

[6] The resolution limit for visible light and an NA = 1 is approximately $r_{\text{res}} = \frac{\lambda}{2\text{NA}} \approx 250$ nm.

and out of the focal plane. Hence they are observed for a short moment only and thus the observed particle trajectories are short, which in turn can lead to poor statistics. Moreover, because their rotational diffusion is also fast, high frame rates would be required not to miss important information between frames. Consequently, this yields low signal to noise ratios for the detection of small scatterers. Also, deducing statistically significant results by tracking individual self-propelling particles and evaluating their MSD is vulnerable to errors emerging from either bad experimental conditions [39] or averaging too few particles [40]. This emphasizes that imaging-based dark-field microscopy is highly problematic due to practical and conceptual difficulties that arise. However, owing to the small size and thereby low sedimentation speeds it was possible to perform scattering experiments on dense colloidal suspensions of Janus particle ensembles. This had the advantage of excellent statistics and because the measurement happened in bulk solution it simultaneously ruled out the aforementioned surface interactions. Therefore the experiments conducted with achiral and highly symmetric Janus particles were based on light scattering of particle ensembles, i.e. Dynamic Light Scattering (DLS), Differential Dynamic Microscopy (DDM) and Super-Heterodyne Laser Doppler Velocimetry (SH-LDV).

Irrespective of the exact propulsion mechanism the active motion of self-propelled Janus particles is definitely generated intrinsic and one of the most commonly employed materials is platinum, which catalytically decomposes H_2O_2 into H_2O and oxygen. Thus, one set of experiments was performed with SiO_2 beads that where half coated with Pt. In the presence of hydrogen peroxide the catalytic activity of Pt generates a concentration gradient of educt and product molecules ∇C. This causes the active motion of the particle. A second material system consisting of TiO_2 coated SiO_2 beads was also investigated [25, 41]. It is also based on the catalytic decomposition of H_2O_2 which causes active propulsion, but here the chemical reaction is triggered by UV-light. In fact, anatase-TiO_2 is a semiconductor with a band gap in the UV and upon illumination with UV light electron-hole pairs are generated, which in turn can drive the decomposition of hydrogen peroxide due to a redox reaction:

$$\begin{array}{c} H_2O_2 + 2e^- + 2H^+ \longrightarrow 2H_2O \quad \text{(Reduction)} \\ H_2O_2 \longrightarrow 2H^+ + O_2 + 2e^- \quad \text{(Oxidation)} \end{array} \tag{3.12}$$

This JP is schematically sketched in Fig. 3.14a and earlier experiments with the TiO_2/SiO_2 system already observed negative phototactic propulsion, i.e. the directional motion away from the UV light source [25, 41].

Fabrication of both, the TiO_2/SiO_2 as well as the Pt/SiO_2 Janus particles was done by using standard PVD (without oblique incidence) to coat a monolayer of silica beads on one side. Beads with various sizes (r_{theo}=72 nm, 130 nm, 177 nm, 259 nm and 575 nm specified by the supplier *microParticles GmbH*) were used to examine the size-dependent propulsion characteristics and an SEM image is shown in Fig. 3.14b. Each monolayer has been prepared by Langmuir–Blodgett deposition as explained in Sect. 2.7. After the deposition, the beads were immersed in Milli-Q water via ultrasonication. Before sonication the TiO_2/SiO_2 particles were annealed to 450°C for

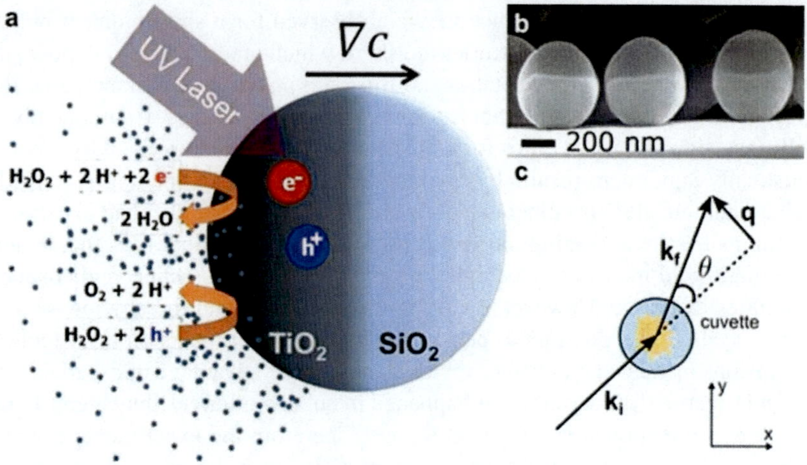

Fig. 3.14 Spectroscopic observation of chemically powered Janus particles. **a** Schematic drawing of a TiO_2/SiO_2 Janus particle which is photocatalytically active upon irradiation with UV light. Figure taken from reference [2]. **b** SEM image of the corresponding Janus particles with radius $r = 259$ nm (specified by the supplier) shows the TiO_2 cap. **c** Schematic drawing of the scattering geometry conducted on a high density particle ensemble in a cuvette, e.g. in a typical DLS experiment. The incident beam $\mathbf{k_i}$ and scattered beam $\mathbf{k_f}$ determine the scattering vector $\mathbf{q}$, which is thus dependent on the scattering angle θ

120 min in order to convert the deposited rutile-TiO_2 into the anatase crystal-phase whose band gap ($E_{\text{gap}} = 3.2$ eV) lies in the UV.

3.6.1 *Spectroscopic Observation of Particle Ensembles*

3.6.1.1 Dynamic Light Scattering

First, DLS experiments to characterize the motion of colloidal Janus particle ensembles were conducted, based on a commercial setup which utilizes a green laser ($\lambda = 532$ nm) for the measurements (*LS Instruments*). The detailed theoretical description of light scattering [42] and in particular the theory of the DLS technique can be found elsewhere [43]. Briefly, a colloidal suspension of particles is illuminated with coherent light, e.g. a laser, and the scattered intensity is recorded by a detector over time $I(t)$. Normally a photo-multiplier tube or an avalanche photodiode is employed to measure the homodyne signal, i.e. the signal which records only light scattered by the particles. The scattered light depends on the scattering angle θ under which it is detected and that corresponds to a scattering vector:

$$\mathbf{q} = \mathbf{k_f} - \mathbf{k}_i = \frac{4\pi n}{\lambda} \sin(\theta / 2), \tag{3.13}$$

where n is the refractive index, and $\mathbf{k_i}$ and $\mathbf{k_f}$ are the wavevectors of the incident and scattered light as schematically depicted in Fig. 3.14c. The detected intensities are subject to interference of the light scattered off by individual particles and hence, if they displace among each other, the interference also changes. Therefore a random motion of the particles leads to characteristic fluctuations in the detected intensity. Intuitively, a faster motion of the colloids leads to a faster fluctuation rate. The fluctuation speed and thereby the characteristic diffusion time is typically observed by calculating an autocorrelation function from the intensity fluctuations, which is subsequently fitted with an appropriate fit model. Thus it is a prerequisite to know the theoretical model describing the particles' motion a priori. The calculation is typically performed on a hardware correlator, which provides a much better time resolution than software based correlation.

However, the standard fit model is only valid for an isotropic motion, e.g. for randomly moving Brownian particles, whereas Janus particles exhibit more complex motion characteristics. Attempts to provide a theoretical framework for JP have been reported [44, 45], but remain specific to the model system and tedious to employ. Especially the directionality of TiO_2/SiO_2 Janus swimmers is not considered by any of the fit models so far. Thus, the idea was to omit the fitting and rather compare raw autocorrelation functions from passive and active Janus particles to get an idea about the diffusion enhancement they exhibit in their active state. For this, a UV laser with $\lambda_{UV} = 375$ nm (*Coherent* OBIS LX375) was incorporated in the DLS setup. Its illumination direction was perpendicular with respect to the scattering plane, here the xy (see Fig. 3.14c) and consequently the laser was parallel to $\hat{\mathbf{z}}$. The raw autocorrelation curves were evaluated by a Matlab script but no reliable results could be obtained from the measured data owing to experimental difficulties, irrespective if TiO_2/SiO_2 or Pt/SiO_2 particles have been examined. The decomposition of H_2O_2 led to bubbling which perturbed the scattered intensities and caused a safety shut down of the detector. Even in case the signal was observable, no systematic behaviour of the autocorrelation function was deduced from repeated measurements. Instead, the signals were more or less random when the passive and active motion was compared. Altogether, the data did not yield meaningful results at all. However, the experiment indicated that the active motion of JP strongly depend on the experimental boundary conditions and that the stability of those parameters is crucial for a comparison of active to passive Janus particle diffusion.

3.6.1.2 Differential Dynamic Microscopy

Next, DDM was used for characterizing the propulsion of Janus particles, which is complementary to DLS. It uses smaller scattering angles and thereby scattering vectors $\mathbf{q}$. Microscope videos of colloidal suspensions are recorded and evaluated to extract the diffusion constant of the particle ensemble. Importantly, it is not necessary to optically resolve the individual scatterers although standard bright-field microscopy is employed. Instead, individual frames of a video are subtracted and a Fourier transformation is applied to access the image structure function which is proportional to the autocorrelation function measured by DLS. Again, a fit model has to be employed in order to deduce the diffusion constant from it. Note, that in DDM measurements the detector is a conventional camera consisting of a pixel array. Thus, one advantage is that not only scattering at a single angle θ but rather many different angels are recorded simultaneously. Similar to DLS the characteristic intensity fluctuation is evaluated, here the fluctuation a single pixel exhibits over time. The method is relatively new and was introduced roughly one decade ago [46]. During this thesis it was implemented in Matlab in order to evaluate the data acquired for passive and active JP (see Appendix A.5).

Initially the code was confirmed to work properly with passive SiO_2 beads as shown in the Appendix (A.5) and afterwards TiO_2/SiO_2 Janus particles were examined. This time they have been activated with a a light emitting diode (*Thorlabs* M365LP1) with center wavelength $\lambda_{UV} = 365$ nm, which was illuminating the particles from the bottom and thereby perpendicular to the microscopes' focal plane. The directionality of the JP led to a propulsion in and out of focus of the microscope. Unfortunately this was not detectable with DDM as the method is limited to motion which is happening in 2D, i.e. motion parallel to the focal plane. However, Pt/SiO_2 particles are predicted to exhibit an isotropic motion and thus DDM is in general capable of measuring their diffusion. But again, no reliable data with a systematic behaviour was observed for repeated measurements on passive and active JP. Other studies showed that in general DDM can be employed to examine Pt-capped and H_2O_2 fueled swimmers [45], utilizing complex fitting models. But this was done with bigger sized particles on a surface. However, the negative results reported here again indicate that stability of experimental boundary parameters is a key prerequisite for the controlled propulsion of chemically powered microswimmers. For JP with small sizes this issue is even more challenging than for the bigger particles.

These observations are in agreement with other reports, which strongly suggest that the chemical environment has a dramatic influence on the propulsion of Janus colloids, and therefore this is regarded as a possible explanation for the inconsistent reports on the swimming velocity of self-propelled Janus particles [31]. The experiments done here exhibit excellent statistics due to an average measurement on high density ensembles and therefore hint that the inconsistency in the literature can also stem from low statistics because experiments are typically performed with few particle tracks leading to unreliable MSDs. Low statistics as well as bad experimental design have both been proposed as problematic for the observation of self-propelling particles [39, 40].

3.6.1.3 Super-Heterodyne Laser Doppler Velocimetry

Finally, SH-LDV has been employed, which is another technique to spectroscopically characaterize the collective motion of particle ensembles. The experiments were undertaken at the *Institute of Physics, Johannes-Gutenberg University Mainz* where Denis Botin supervised and operated the setup. Laser Doppler Velocimetry is again closely related to DLS as it measures the Fourier transform of the intensity autocorrelation function. Hence, a similar setup as shown in Fig. 3.14c was utilized, whereas the measurement wavelength was $\lambda = 633$ nm in this instrument. However, now the detected intensity fluctuations were observed in the frequency domain by using a hardware frequency analyzer (*OnoSokki* DS2000) instead of the hardware correlator behind the detector. Thus the power spectrum $C(\mathbf{q}, \omega)$ of the intensity time trace $I(t)$ is recorded. Moreover, the technique was combined with super-heterodyning, i.e. the detected intensity was overlapped with an artificial reference beam (see Fig. 3.15a) of known frequency.[7] Consequently an artificial beat is modulated on top of the characteristic intensity fluctuations and the signal-to-noise ratio is significantly increased similar to lock-in amplification where an external frequency is modulated to obtain the measurement signal at a desired frequency. However, the main advantage of SH-LDV was that interpretation was significantly simplified as it does not require fitting and thus no underlying propulsion mechanism has to be presumed a priori. Instead the velocity distribution of the particles in the scattering volume can be obtained without the need for a model. Additionally, the instrument provided easy access for a UV laser, which triggered the photocatalytic decomposition of TiO_2 parallel to the scattering plane, i.e. the $x\,z$, which permitted the incident direction to be switched ($\mathbf{k}_{UV}$). A sketch of the experimental geometry is shown in Fig. 3.15a. To drive the photocatalytically active JP a laser (*Thorlabs* L375P70MLD) that emits at ($\lambda_{UV} = 375$ nm) has been used.

A comprehensive theory of SH-LDV is beyond the scope of this thesis and can be found elsewhere [2, 47–49]. Importantly, the homodyne and heterodyne signals are well-separated in frequency space by the frequency difference between the illumination and reference beam. This difference is also called the Bragg frequency and here it was $\omega_{Bragg} = 3000 \pm 0.2$ Hz. Thus, for passive Brownian particles a Lorentzian was observed centered around ω_{Bragg}, as shown in Fig. 3.15b for TiO_2/SiO_2 particles ($r_{theo} = 259$ nm) in 2.5% H_2O_2. However, upon irradiation with UV light, the particles started to actively move. This led to an in- or decrease of the recorded frequency because the particles changed their position with respect to the detector and thus the frequency of the scattered light appears slightly shifted due to the Doppler effect (Fig. 3.15b). In fact, the setup was sensitive to the velocity parallel to the scattering vector $\mathbf{q}$, which is (in this setup) equal to measuring the projection of the swimming velocity onto the z-axis. A control measurement with the same particles immersed in pure water revealed that the Lorentzian around ω_{Bragg} did not change its position even when the sample was illuminated with UV light. Hence, the Janus particles indeed remained passive without fuel.

Interestingly, the Doppler effect yields an increase or a decrease of the observed frequency depending on whether the source moves towards or away from the detector, respectively. This implies that a motion parallel or anti-parallel to $\mathbf{q}$ leads to a positive or negative frequency shift in the SH-LDV spectrum. Indeed, this was demonstrated for the same particles in Fig. 3.15c which shows the shifted frequency distribution upon illumination of the sample with an angle of incidence of $\alpha = \pm 18.8°$ between

[7] In practice the artificial reference beam was realized by shifting the frequency of the illumination beam with an acousto-optical modulator (*APE GmbH*) by a desired Bragg frequency ω_{Bragg}.

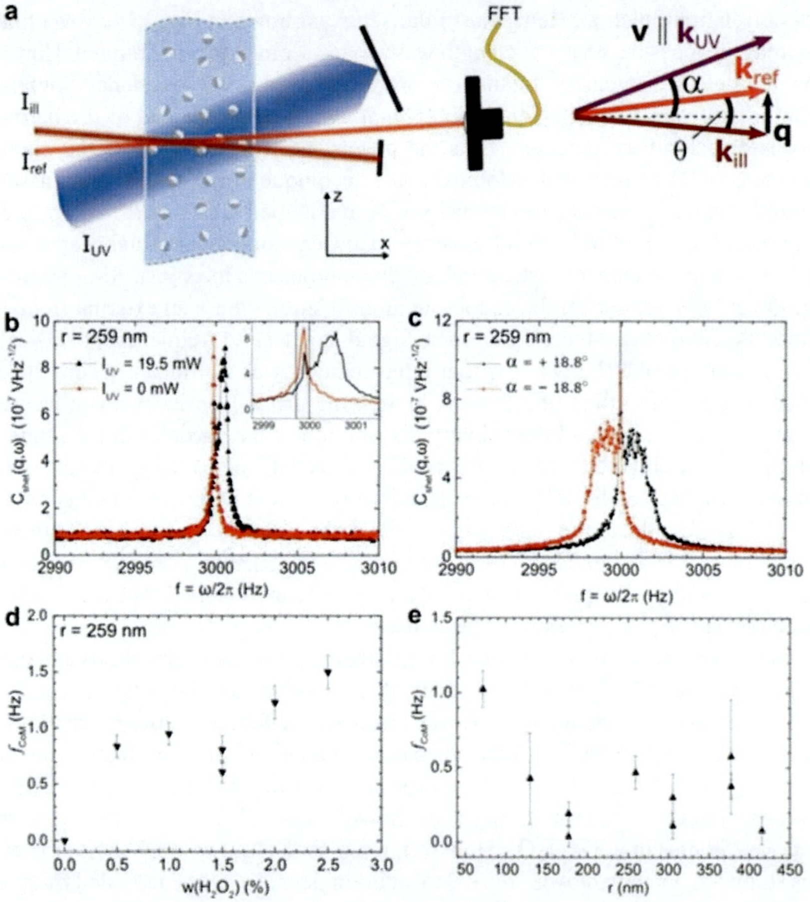

Fig. 3.15 Measurement of TiO_2/SiO_2 Janus particles ($r_{\text{theo}} = 259$ nm) with Super-Heterodyne Laser Doppler Velocimetry (SH-LDV). **a** Setup where the illumination ($\mathbf{k_{ill}}$) and reference ($\mathbf{k_{ref}}$) beams cross in the sample cell causing a heterodyne signal. The UV light sources' incident angle α was variable. **b** Passive Brownian particles result in a Lorentzian centered around $\omega_{\text{Bragg}} =$ 3000 Hz, whereas a broadened and Doppler-shifted signal is found for actively moving particles (inset shows zoom around ω_{Bragg}). **c** A change in the incident direction of the UV light results in a Doppler shift with opposite sign owing to a reversed swimming direction (projected on the z-axis). **d** The particles' center-of-mass frequency shift (i.e. velocity) was found to increase upon increasing H_2O_2 concentration. **e** Size-dependent swimming seemingly shows an increased frequency shift for smaller particles, but also large variations between the individual experiments that indicate non-homogeneous boundary conditions. Figure adapted from Ref. [2]

$\mathbf{k_{UV}}$ and the recorded intensity $\mathbf{k_{ref}}$ (see Fig. 3.15a). The Janus particles move away from the light source in both situations which confirms their phototactic behaviour in 3D.

In Fig. 3.15b, c it is also seen that the active motion leads not only to a shifted frequency but sometimes also to a broadening of the shifted signals. This broad peak represents a distribution of velocities that is measured which may arise due to different sizes, different amount of photocatalytically active TiO_2, or lower UV light intensities at the edge of the UV laser beam that may lead to minute variations in the velocity individual particles exhibit. Because SH-LDV is a highly sensitive technique, it will resolve such differences and display a broad distribution of Doppler shifted velocities. To compare different measurements the center-of-mass frequency of this velocity distribution was used, which is equal to the ensemble-averaged particle velocity. Note, that an isotropic motion would only lead to a symmetric broadening around the Bragg frequency but the center of mass will not be Doppler shifted.

Another advantage of the SH-LDV technique was that the velocity distribution could be monitored over time and hence artificial perturbations become readily observable, i.e. a stable active motion can be distinguished from signals that are unstable and vary with time [2]. Therefore, it is especially useful for the examination of varying experimental boundary conditions, which were already identified to play a crucial role for the self-propulsion of a JP. To see the dependency of the JP velocity on the amount of hydrogen peroxide a series of experiments with different dilutions of H_2O_2 was carried out. The corresponding center of mass frequencies are shown in Fig. 3.15d. As described, the measurement in pure water confirmed that in the absence of fuel no active propulsion occurs. If the concentration of H_2O_2 increases the swimming velocity has the tendency to increase linearly. This is in agreement with other studies on bigger particles [25, 26]. However, although each individual experiment reliably determined the center of mass frequency the overall data exhibits outliers and a functional dependency could not be deduced. Given the fact that SH-LDV provides excellent statistical significance, this suggests that even minute variations during sample preparation can have a strong impacts on the active propulsion. Another series of experiments was undertaken with differently sized Janus particles (Fig. 3.15e) to see how the swimming velocity scales with decreasing the size of the JP. Knowledge of the size-dependent diffusion characteristics would help to clarify the underlying propulsion mechanism [35]. It was observed that qualitatively the frequency shift, i.e. the swimming speed, has the tendency to increase. However, repeated experiments again showed large variation and thus dramatic differences for one and the same set of boundary conditions, in agreement with the DLS and DDM experiments. Again, an individual SH-LDV measurement was obtained with high statistical confidence, but the overall results appear uncertain and ultimately indicate that imperfections during fabrication of the differently sized Janus particles or any other experimental parameter will crucially impact the propulsion characteristics. A final assessment about the underlying propulsion mechanism was thus not possible.

Nevertheless, in contrast to DLS and DDM, the SH-LDV technique is a valuable tool to determine swimming velocity distributions and is at the same time sensitive to the swimming direction. Therefore, the directionality of the active motion exhibited by TiO_2/SiO_2 Janus particles was unequivocally demonstrated and preliminary results for varying boundary conditions are presented. The technique was identified as suitable for the examination of self-propelled Janus particles because of its excellent statistics and because it permits a model-free interpretation of the data. Most importantly, it yields the possibility to recognize insufficiently stabilized environmental parameters, which clearly is important given the fact that other experimental studies to date yield inconsistent and contradictory results for similar Janus swimmers.

3.7 Conclusions and Outlook

In conclusion, the analytical theory, which has been published [19] and which describes the rotation-translation coupling at low Reynolds number of an arbitrary shaped magnetic object was experimentally proven by the work presented in this chapter. First, a cm-sized macroscopic model of an arc was used to fix the orientation of the magnetic dipole moment with respect to the shape and to demonstrate the dependence of the propulsion characteristics on the orientation of the dipole moment. Second, V-shapes grown by PVD were used to validate the applicability of the theory to other structures that are only a few microns in size. Additionally, the microstructures were not exclusively actuated with a magnetic field, instead an electric field has been used as well. The latter was employed to demonstrate the first achiral object that propels due to rotation-translation coupling at low Reynolds number.

Based on the findings and alongside the experiments the existing theory was extended in order to provide symmetry arguments that predict the characteristic non-, bi- or unidirectional propulsion of highly symmetrical shapes. The presented work highlights the importance to include also the dipole moment into a rigorous analysis of the symmetries, rather then just the shape alone. Although this was already known (see Sect. 2.5.1 and [18]) it has been overlooked in recent studies [17]. Surprisingly, an achiral shape like the "V" together with an achiral dipole moment $\boldsymbol{m}$ can render an object to be chiral. Those objects are shown to be propulsive, which is, however, not surprising because a chiral-shaped corkscrew efficiently converts rotation into a linear motion, too. In fact, the chiral ($\widehat{P}$-odd) V-shapes exhibit the same unidirectional motion identical to a ($\widehat{P}$-odd) corkscrew. In contrast, it was demonstrated that objects with higher symmetry ($\widehat{C}\widehat{P}$-even for magnetically driven and $\widehat{P}$-even for electrically driven) can also be propulsive. These findings can have important practical implications. They suggest a novel and much simpler design for wirelessly powered micro- and nanoswimmers, which are currently at the focus of ongoing highly interdisciplinary research efforts.

In the introduction to this chapter the famous words of E. Purcell were mentioned: "Turn anything—if it isn't perfectly symmetrical, you'll swim". His statement can now be modified to read: "Turn anything around any axis which is not a principal axis of rotation—even if it's highly symmetrical, you'll swim". Both statements imply the necessity to break the time-reversal symmetry of the Stokes-equation (Eq. (2.5)) for efficient motion at low Re. In future, it will be interesting to further examine the symmetries and ability of highly symmetrical and externally driven objects to propel. One idea would be to use a conical external magnetic field, which is a superposition of a rotating field and a constant field (the constant component is along the rotations' angular velocity vector and the field vector thus precess around this axis). Then, the symmetry of the problem would be broken due to the external field and also magnetic objects of highest symmetry, e.g. $\widehat{C}\widehat{P}$- and $\widehat{P}$-even objects, would be propulsive. In fact, the V-shapes that have their moment aligned parallel to one of the principal axes of rotation could then exhibit a translational motion, too.

Even simpler shapes than the "V" can also generate propulsion at low Re although for spherical shapes rotation-translation coupling and thus physically powered propulsion fails. But spherical and chemically powered Janus particles can still exhibit propulsion. Their smallest size approaches a few nanometers and then scattering based techniques need to been employed to investigate their motion. Of particular interest are dense suspensions. Methods like DLS and DDM failed to measure reliable results and moreover suffer from the fact that a fitting model is required a priori. This means a propulsion mechanism has to be presumed to interpret the experiments. However, the underlying mechanism for self-propulsion of JP is not yet fully resolved. Conversely, SH-LDV experiments were identified as a suitable method for the characterization of the collective motion exhibited by Janus particles, as it is model-free and provides rich information on the velocity distribution and direction. The directionality of self-propelled TiO_2/SiO_2 particles was confirmed and altogether the stability of experimental boundary conditions was shown to be a key pre-requisite for the examination of such systems. Preliminary results indicate an increase in propulsion speed with decreasing particle size. Nevertheless, the results are difficult to reproduce quantitatively. Therefore, it is necessary to conduct further experiments on those systems in future. This could pave the way to ultimately clarify the exact mechanism which causes Janus *nano*particles to move.

References

1. Sachs J, Morozov KI, Kenneth O, Qiu T, Segreto N, Fischer P, Leshansky AM (2018) Role of symmetry in driven propulsion at low Reynolds number. Phys Rev E 98(6)
2. Sachs J, Kottapalli SN, Fischer P, Botin D, Palberg T (2021) Characterization of active matter in dense suspensions with heterodyne laser doppler velocimetry. Colloid Polym Sci 299:269–280
3. Nelson BJ, Kaliakatsos IK, Abbott JJ (2010) Microrobots for minimally invasive medicine. Annu Rev Biomed Eng 12:55–85
4. Purcell EM (1977) Life at low Reynolds number. Am J Phys 45:3
5. Shah AS, Ben-Shahar Y, Moninger TO, Kline JN, Welsh MJ (2009) Motile cilia of human airway epithelia are chemosensory. Science 325(5944):1131–1134
6. Lauga E, Powers TR (2009) The hydrodynamics of swimming microorganisms. Rep Prog Phys 72(9):096601
7. Guo S, Sawamoto J, Pan Q (2005) A novel type of microrobot for biomedical application. In: 2005 IEEE/RSJ international conference on intelligent robots and systems. IEEE, pp 1047–1052

8. Berg HC, Anderson RA (1973) Bacteria swim by rotating their flagellar filaments. Nature 245(5425):380–382
9. Bell DJ, Leutenegger S, Hammar KM, Dong LX, Nelson BJ (2007) Flagella-like propulsion for microrobots using a nanocoil and a rotating electromagnetic field. In: Proceedings 2007 IEEE international conference on robotics and automation. IEEE, pp 1128–1133
10. Ghosh A, Fischer P (2009) Controlled propulsion of artificial magnetic nanostructured propellers. Nano Lett 9(6):2243–2245
11. Walker D, Kubler M, Morozov KI, Fischer P, Leshansky AM (2015) Optimal length of low Reynolds number nanopropellers. Nano Lett 15(7):4412–4416
12. Schamel D, Mark AG, Gibbs JG, Miksch C, Morozov KI, Leshansky AM, Fischer P (2014) Nanopropellers and their actuation in complex viscoelastic media. ACS Nano 8(9):8794–8801
13. Wu Z, Troll J, Jeong H-H, Wei Q, Stang M, Ziemssen F, Wang Z, Dong M, Schnichels S, Qiu T, Fischer P (2018) A swarm of slippery micropropellers penetrates the vitreous body of the eye. Sci Adv 4(11):eaat4388
14. Peyer KE, Zhang L, Nelson BJ (2013) Bio-inspired magnetic swimming microrobots for biomedical applications. Nanoscale 5(4):1259–1272
15. Vach PJ, Brun N, Bennet M, Bertinetti L, Widdrat M, Baumgartner J, Klumpp S, Fratzl P, Faivre D (2013) Selecting for function: solution synthesis of magnetic nanopropellers. Nano Lett 13(11):5373–5378
16. Vach PJ, Fratzl P, Klumpp S, Faivre D (2015) Fast magnetic micropropellers with random shapes. Nano Lett 15(10):7064–7070
17. Cheang UK, Meshkati F, Kim D, Kim MJ, Fu HC (2014) Minimal geometric requirements for micropropulsion via magnetic rotation. Phys Rev E 90(3):033007
18. Barron LD (1983) Molecular light scattering and optical activity. Cambridge University Press, Cambridge
19. Morozov KI, Mirzae Y, Kenneth O, Leshansky AM (2017) Dynamics of arbitrary shaped propellers driven by a rotating magnetic field. Phys Rev Fluids 2(4):044202
20. Happel J, Brenner H (1983) Low Reynolds number hydrodynamics with special applications to particulate media. Monographs and textbooks on mechanics of solids and fluids, 1st paperback edn. Martinus Nijhoff, The Hague
21. Fan DL, Zhu FQ, Cammarata RC, Chien CL (2005) Controllable high-speed rotation of nanowires. Phys Rev Lett 94(24):247208
22. Sánchez S, Soler L, Katuri J (2015) Chemically powered micro-and nanomotors. Angew Chem Int Ed 54(5):1414–1444
23. Dey KK, Wong F, Altemose A, Sen A (2016) Catalytic motors-quo vadimus? Curr Opin Colloid Interface Sci 21:4–13
24. Brown AT, Poon W (2014) Ionic effects in self-propelled pt-coated janus swimmers. Soft Matter 10(22):4016–4027
25. Singh DP, Choudhury U, Fischer P, Mark AG (2017) Non-equilibrium assembly of light-activated colloidal mixtures. Adv Mater 29(32):1701328
26. Wheat PM, Marine NA, Moran JL, Posner JD (2010) Rapid fabrication of bimetallic spherical motors. Langmuir 26(16):13052–13055
27. Lee T-C, Alarcon-Correa M, Miksch C, Hahn K, Gibbs JG, Fischer P (2014) Self-propelling nanomotors in the presence of strong brownian forces. Nano Lett 14(5):2407–2412
28. Kline TR, Paxton WF, Mallouk TE, Sen A (2005) Catalytic nanomotors: remote-controlled autonomous movement of striped metallic nanorods. Angew Chem Int Ed 44(5):744–746
29. Walker D, Käsdorf BT, Jeong H-H, Lieleg O, Fischer P (2015) Enzymatically active biomimetic micropropellers for the penetration of mucin gels. Sci Adv 1(11):e1500501
30. Howse JR, Jones RAL, Ryan AJ, Gough T, Vafabakhsh R, Golestanian R (2007) Self-motile colloidal particles: from directed propulsion to random walk. Phys Rev Lett 99(4):048102
31. Ketzetzi S, de Graaf J, Doherty RP, Kraft DJ (2020) Slip length dependent propulsion speed of catalytic colloidal swimmers near walls. Phys Rev Lett 124(4):048002
32. Crowdy DG (2013) Wall effects on self-diffusiophoretic janus particles: a theoretical study. J Fluid Mech 735:473–498

33. Uspal WE, Popescu MN, Dietrich S, Tasinkevych M (2015) Self-propulsion of a catalytically active particle near a planar wall: from reflection to sliding and hovering. Soft Matter 11(3):434–438
34. Brooks AM, Tasinkevych M, Sabrina S, Velegol D, Sen A, Bishop KJM (2019) Shape-directed rotation of homogeneous micromotors via catalytic self-electrophoresis. Nat Commun 10(1):1–9
35. Brown AT, Poon WCK, Holm C, de Graaf J (2017) Ionic screening and dissociation are crucial for understanding chemical self-propulsion in polar solvents. Soft Matter 13(6):1200–1222
36. Wei M, Zhou C, Tang J, Wang W (2018) Catalytic micromotors moving near polyelectrolyte-modified substrates: the roles of surface charges, morphology, and released ions. ACS Appl Mater Interfaces 10(3):2249–2252
37. Palacci J, Sacanna S, Kim S-H, Yi G-R, Pine DJ, Chaikin PM (2014) Light-activated self-propelled colloids. Philos Trans R Soc A: Math, Phys Eng Sci 372(2029): 20130372
38. Leeth Holterhoff A, Li M, Gibbs JG (2018) Self-phoretic microswimmers propel at speeds dependent upon an adjacent surface's physicochemical properties. J Phys Chem Lett 9(17):5023–5028
39. Dunderdale G, Ebbens S, Fairclough P, Howse J (2012) Importance of particle tracking and calculating the mean-squared displacement in distinguishing nanopropulsion from other processes. Langmuir 28(30):10997–11006
40. Novotný F, Pumera M (2019) Nanomotor tracking experiments at the edge of reproducibility. Sci Rep 9(1):1–11
41. Singh DP, Uspal WE, Popescu MN, Wilson LG, Fischer P (2018) Photogravitactic microswimmers. Adv Func Mater 28(25):1706660
42. Berne BJ, Pecora R (2000) Dynamic light scattering: with applications to chemistry, biology, and physics. Courier Corporation
43. Stetefeld J, McKenna SA, Patel TR (2016) Dynamic light scattering: a practical guide and applications in biomedical sciences. Biophys Rev 8(4):409–427
44. Kurzthaler C, Leitmann S, Franosch T (2016) Intermediate scattering function of an anisotropic active brownian particle. Sci Rep 6:36702
45. Kurzthaler C, Devailly C, Arlt J, Franosch T, Poon WCK, Martinez VA, Brown AT (2018) Probing the spatiotemporal dynamics of catalytic janus particles with single-particle tracking and differential dynamic microscopy. Phys Rev Lett 121(7):078001
46. Cerbino R, Trappe V (2008) Differential dynamic microscopy: probing wave vector dependent dynamics with a microscope. Phys Rev Lett 100(18):188102
47. Drain LE et al (1980) The laser Doppler technique. Wiley, New York
48. Adrian RJ (1993) Selected papers on laser Doppler velocimetry, vol 78. Society of Photo Optical
49. Botin D, Wenzl J, Niu R, Palberg T (2018) Colloidal electro-phoresis in the presence of symmetric and asymmetric electro-osmotic flow. Soft Matter 14(40):8191–8204

Chapter 4
Chiroptical Spectroscopy of Single Chiral and Achiral Nanoparticles

This chapter is based on and contains excerpts and figures from the article "Chiroptical Spectroscopy of a Freely Diffusing Single Nanoparticle" [1]. Contributions of coauthors are indicated.

4.1 General Motivation

Chiroptical spectroscopy is the analysis of differences that occur between the intensities of left- and right-circularly polarized light and is therefore also denoted as circular differential intensity. It is of utmost importance for natural sciences and life in general because almost every biomolecule and drug is chiral. Their handedness and purity are typically observed by optical spectroscopy, which is based on a difference between left- and right-circularly polarized light (LCP, respectively RCP) interacting with a bulk solution containing a chiral analyte. The strength of this interaction is rather weak because it relies on a coupling of magnetic and electric dipoles of a single molecule [2]. However, by probing ensembles of many molecules the individual signals add up and can be detected, e.g. with typical circular dichroism (CD) spectroscopy measuring the attenuation of LCP and RCP light of a substance. In practice either high concentrations or long path lengths are necessary to detect chirality in the molecular samples. A prominent example of the capabilities of classical CD spectroscopy is the power to resolve the secondary structure of proteins [3]. Traditional spectrometers measure the CD of a molecular ensemble in solution at one wavelength at a time by modulating the polarization with a pockels-cell and a lock-in amplification scheme [4, 5]. Nowadays, the pockels cell is usually replaced by photo-elastic modulators, that still temporally modulate the input light's polarization state. A drawback of this approach is that it requires relatively large sample volumes. In addition, the ensemble measurement also causes a loss of important

J. Sachs, *Motion, Symmetry & Spectroscopy of Chiral Nanostructures*, Springer Theses,
https://doi.org/10.1007/978-3-030-88689-9_4

spectral information [6]. Especially information on dynamic processes and important variations within the analyte distribution (which is particularly relevant in the case of nanoparticles) can be lost by ensemble averaging, highlighting the relevance for single molecule and single-particle experiments.

Measurements of few nanostructures and in some cases individual nanostructures have been reported. However, as will be discussed later, these measurements are not identical to the traditional ensemble spectroscopies, as they are performed on immobilized particles or structures, where artefacts can arise. Nevertheless, enhancing the weak chiroptical signals with the help of plasmonic nanostructures has become a major research focus in the recent decade. The proposed existence of "superchiral" light in nanostructures [7] was followed by several experimental realizations that claimed a dramatic enhancement in the magnitude of chiral intensity differences. Most prominently, Kadodwala et al. used gold gammadion structures and reported an increase of 10^6 of the relevant observable [8]. Later, Alu et al. claimed zeptoliter ($10^{-21}l$) sensitivities for chiral molecules based on plasmonic structures [9]. Such an enhancement would open up the possibility of detection down to single molecules, which also dramatically reduces the necessary sample volumes compared to traditional CD spectroscopy. It is important to note that there are severe practical and conceptual difficulties with plasmonic nanostructures, if they are probed on a surface [10], rendering many of the experimental reports to date highly debatable. The reason is that plasmonic nanoparticles of basically any shape (including achiral structures) can interact differently with left- and right-circularly polarized light. Immobilization on a surface will fix the nanostructures in 3D-space and thus orient them with a specific alignment with respect to the incident electric field vectors. A strong difference in the interaction with left- and right-circularly polarized light can arise, but the origin of these chiroptical signals are linear optical anisotropies, i.e. linear dichroism and linear birefringence (and not structural chirality) [10, 11], and consequently they are not a measure of the chirality intrinsic to the nanostructures. Interpretation of chiroptical signals acquired for a fixed light-object geometry are therefore complicated. This has long since been realized by chemists [12–14], especially because orientation has led to severe misinterpretation of molecular CD spectra in the past [15]. Plasmonic nanoparticles are likely to suffer from the same difficulties and it is therefore highly desirable to have a robust experimental approach that is easy to interpret and that minimizes artefacts—something that has been achieved in this thesis and that is reported in this chapter.

The theory of molecular optical activity shows that in the case of a fixed orientation of molecules, such as molecules at an interface, the optical activity contains contributions not only from the optical activity tensor (the mixed electric-magnetic dipole polarizability), but also from electric-quadrupolar terms (that also exist for achiral molecules) [12]. In contrast, when measuring a bulk ensemble of isotropically oriented molecules, the quadrupolar terms will cancel and only the isotropic component of the optical activity tensor, a pseudoscalar, survives orientational averaging. Because the circular birefringence (CB) is related to the circular dichroism (CD) by Kramers-Kronig relation, it is not surprising that classical CD spectroscopy can produce the same artefacts, if the sample is not isotropic. Additionally, it has been proven that effects due to birefringence and dichroism of linear polarizations (LB and LD) can also contribute to spectra of molecules at interfaces [14]. Similar arguments can be employed for modelling plasmonic nanostructures where several electric (and magnetic) dipoles and other multipoles are excited, which interact with the incident light field as well as the field of the other dipoles [10, 16–18].

The vast majority of previous spectroscopic studies considering chiral plasmonics involve surfaces and thus run into the same difficulties as oriented molecular systems. Even more problems arise because of the quasi-2D geometry of such structures, which means they are not exactly 2-dimensional (which is the symmetry that is assumed), but rather they extend a few nano- or micrometers into the third dimension. This complicates the interpretation of chiroptical measurements [11]. Unfortunately, most of the reports regarding chiro-plasmonic nano- and microstructures pay little attention to influences that appear as a circular differential intensity difference, but do not reflect chirality intrinsic to the sample being probed. The reports thus lack significance unless comprehensive control measurements are executed [10], or they require sophisticated evaluation and correction workflows [19, 20] in order to provide an accurate interpretation of the spectra. Some studies rely on careful sample design to partially mitigate the impact of linear polarization effects [21–23]. However, such studies are then limited to highly symmetrical arrays of nanostructures (C_2 or C_4) on a surface where there is also electromagnetic coupling between many individual nanoparticles. Most notably, no work has been published to date that measures the intrinsic chirality of a single nanoparticle. It is thus highly desirable to develop an experimental approach to deduce the true chiroptical spectrum of a single nanoparticle and thereby simplify the spectral interpretation considerably. Prerequisite for such an experiment is a single nano-object, which possesses true geometric chirality and that can be observed while it is freely moving. It can then unhamperedly diffuse, i.e. it has the freedom to reorient spatially while being far away from any surface. Whereas untethered achiral nanorods have been utilized as orientation sensors with the help of linear polarization spectroscopy [24], often in order to perform microrheology [25, 26], there is no report to date on the observation of a single chiral nanoparticle. There is one study that spectroscopically observed chiral nano-cone structures, but it was based on an optical tweezers [27], and so the structure was not freely moving. It is known that the gradient and scattering forces of an optical trap lead to an alignment of shape-anisotropic particles with respect of the incident polarizations [28]. Thus, it can be expected that nano-cones will also align, or at least their orientation is not isotropically sampled during a measurement with the reported setup. Additionally, the investigation was limited to one wavelength only, i.e. no spectrum was acquired, and moreover, it required averaging over a large collection of such particles in order to yield a significant result.

During this thesis a novel single-particle spectroscopy setup was realized and used to study the chiroptical response of differently shaped single nanostructures. The design of the optical layout and the construction of the setup was done by the author together with Jan-Philipp Guenther and Andrew Mark from the *Micro-, Nano- and Molecular Systems* group at the *Max Planck Institute for Intelligent Systems, Stuttgart*. The setup is based on a dark-field microscope and enabled the first true chiroptical spectrum of a single nanoparticle to be acquired in bulk solution, away from a surface. The opportunity to watch a single nanoparticle while it undergoes Brownian motion and at the same time record its differential interaction with circularly polarized light opened the possibility for acquiring a time-series of CDSI spectra. In what follows, it is shown that in order to unequivocally determine the chirality of a nanoparticle, it is necessary to average its spectral response over time (while it reorients). Only a sufficiently complete sampling of all orientations in space allows the true chiral spectral response to be deduced, while at arbitrary instants of time the spectra are not a true measure of chirality. This is especially noteworthy as even achiral nanoparticles, like rods, can exhibit circular intensity differences when oriented, whereas time-averaging leads to the expected vanishing of the CDSI signals for achiral structures. Further, it has been possible to verify the ergodic principle for chiral spectroscopies by demonstrating the equality between a single particles' CDSI spectrum and the corresponding bulk ensemble spectrum. The novel method reported herein has thus all the benefits of chiral bulk spectroscopy and maps these to the observation of one single nanostructure. Thereby a new observable, which has not yet been accessible before, was recorded—the true chiral spectral signature of a single nanoparticle.

The investigated samples were either fabricated with the GLAD (Sect. 2.7) system or were commercially bought gold nanoparticles. The latter are chemically synthesized, giving spectrally narrow plasmon resonances and hence they are suitable as a reference sample. On the other hand, the former provided more complex shapes, such as chiral Au nanohelices, as well as magneto-plasmonic particles. Magnetic fields were used to externally control the alignment of the single particle and thus validate the strong orientation-dependence of the CDSI signals. The rigorous understanding of the chiroptical response of artificial chiral nanostructures is a substantial contribution to the field of chiral plasmonics and it can thus facilitate the development of novel sensing applications, as well as new spectroscopic probes of chirality.

4.2 Novel Dark-Field Spectroscopy Setup

Dark-field microscopy is used to investigate single particles spectroscopically [29, 30], and in particular to explore the different interactions with circularly polarized light [19, 20, 23, 31]. Usually the spectrum is acquired by two independent measurements of the specimen, one with RCP and the other with LCP as incident light. The responses are detected separately and in order to calculate the Circular Differential Scattering Intensity (CDSI). A prerequisite is that they are normalized and background corrected to account for slight intensity fluctuations of the light source between the two measurements. This approach allows only stationary objects to be studied, such that all the particles have a fixed orientation in 3D space. Consequently, difficulties with the spectral interpretation arise, because there can be chirality effects not only due to the geometry of the scatterer, but because of a chiral geometry between the light and the scatterer. The measurement thus records a superposition of circular and linear polarization effects. Complicated correction schemes [19, 20] are employed to mitigate the impact of linear anisotropies, but they necessitate the knowledge of the particle orientation with respect to the input lights' electric field vectors, which is challenging. In contrast, the setup built during this thesis can be used to watch freely diffusing and thus dynamically reorienting particles. Time-resolved chiroptical scattering spectra of single nanoparticles were observed by utilizing a balanced detection scheme, which means both circular polarizations of the scattered light were simultaneously recorded over the visible and near-infrared spectral range. Therefore no complicated correction schemes are necessary, or any a priori knowledge of the particle's orientation. The investigation of single particles that move randomly due to Brownian motion became possible for the first time with this approach. A similar setup has been employed elsewhere [27], but it requires an optical trap to hold particles. The setup here is much simpler as it does not require any complicated trapping optics but rather two quarter-wave plates and a Wollaston prism. Its main advantage is the combination of the balanced detection of LCP and RCP with the ability to spectrally resolve the detected signals. Consequently the information deduced is much richer in the present approach.

Generally, a QWP is used to convert linearly into circularly polarized light and *vice versa*. This is realized by introducing a phase shift of $\frac{\pi}{2}$ between an ordinary and an extraordinary beam, which both propagate through the waveplate and which superimpose behind. The sign of the phase-shift determines whether LCP or RCP is generated, and hence the orientation of the waveplate with respect to the optical axis is important. Alternatively, a $\frac{\pi}{2}$ phase-shift can also be generated with a Pockels-cell or photoelastic modulators, which are used in classical CD spectrometers. They are driven electronically and the handedness of the input polarization can be modulated with high frequency. Thereby a lock-in detection is employed to measure weak chiroptical responses of the output light. However, this approach requires a seperate measurement for every wavelength, which is not practical for the observation of dynamically moving single particles. Therefore, in the present setup the input light is unpolarized and fixed optical elements are used in order to evaluate the polarization state of the output light. The setup thus makes it possible to record a differential spectrum in "one shot".

The core of the setup is a microscope which is followed by relatively simple polarization optics, as schematically depicted in Fig. 4.1. A commercial inverted microscope (Axio Observer, *Zeiss*) has been equipped with a dark-field condenser, which illuminates the sample with an annular light cone of unpolarized light. It has a NA between 1.2 and 1.4. A $40\times$ (NA 0.9) EC Plan-Neofluar objective collected the scattered light before it is redirected via a side-port onto a relay lens system (constructed of two $f = 100$ mm achromatic lenses (NT49374-INK, *Edmund Optics*) to form a parallel beam, such that additional polarization optics can be accommodated. A combination of two superachromatic QWPs (SAQWP05M-700, *Thorlabs*) and a 1° Wollaston prism (WPQ10, *Thorlabs*) were used to spatially separate the RCP and LCP components of the scattered light (Fig. 4.1). First, a QWP at +45° converted CP light into orthogonal linear polarizations that were separated by the following prism (rotated to 0°). After the prism a second QWP was placed at −45°

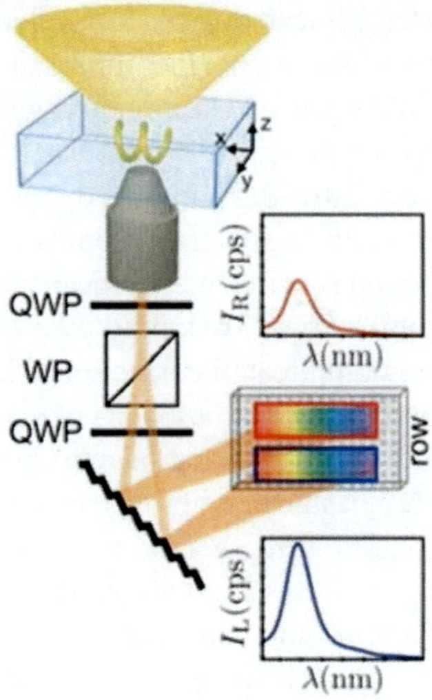

Fig. 4.1 Schematic of the novel single-particle dark-field scattering spectrometer. The incident light is unpolarized. Due to the condenser geometry only light scattered off the nanoparticle is collected by the objective. After exiting the microscope the light is sent through polarization optics consisting of a quarter waveplate (QWP) at +45°, a Wollaston prism (WP) at 0° and a second QWP a −45° before the light is dispersed by a spectrograph and detected on a CCD, where the two independent scattering spectra corresponding to LCP and RCP light are simultaneously recorded. Figure taken from Ref. [1]

that converted the LP light back to circularly polarized. Converting the LP back to circular is crucial to avoid artificial intensity differences that are caused by different Fresnel reflection coefficients for (linear) s- and p-polarizations. The second relay lens focused the two beams onto a Czerny-Turner spectrograph (Shamrock 193i, *Andor*) which was equipped with a 300 lines/mm silver-coated grating and a motorized slit. After the light was dispersed and spectroscopically resolved, it was collected with a CCD detector (DU920P-OE, *Andor*) that has 255 × 1024 pixels, each $26 \times 26\,\mu m^2$ in size. The CCD was an open-electrode type to avoid fringing effects for near-infrared wavelengths and together with the spectrograph has a wavelength resolution of 0.415 nm/px. In front of the relay lens system an adjustable circular aperture (SM1D12D, *Thorlabs*) was introduced to minimize the field of view on the detector and thus cancel out cross-talk between the spatially separate beams. The detection scheme was based on polarization optics that remained fixed, and consequently it was highly important to align the components properly with respect to each other. Otherwise the conversion of LP and CP light can lead to unwanted intensity modulations which can be misinterpreted as chiroptical signals. Spectral and polarization calibration of the setup was done by using 1.5 μm TiO_2 spheres as reference samples, since they are expected to exhibit a spectrally flat scattering signal.

4.2.1 Mueller Matrix Analysis of the Spectrometer

By utilizing the Mueller–Stokes formalism the polarization evolution in the setup can be mathematically described. A brief overview over the formalism and the individual Mueller matrices being used are given in Appendix A.1. While the two QWPs, when properly oriented, convert the CP into LP and *vice versa*, the prism spatially separates the two linear polarization components. The optical train is modelled as two individual channels. The Wollaston prism is represented as a linear polarizer at 0° for channel one and as a polarizer at 90° for the second channel. In doing so, the resulting Mueller matrices for the two channels, corresponding to the spatially separated intensities detected by the CCD are given by:

$$\begin{aligned}\mathbf{M}_{\mathrm{L}} &= \mathbf{M}_{\mathbf{QWP}}(-\mathbf{45}^{\circ}) \cdot \mathbf{M}_{\mathbf{LP}}(\mathbf{0}^{\circ}) \cdot \mathbf{M}_{\mathbf{QWP}}(+\mathbf{45}^{\circ}),\\ \mathbf{M}_{\mathrm{R}} &= \mathbf{M}_{\mathbf{QWP}}(-\mathbf{45}^{\circ}) \cdot \mathbf{M}_{\mathbf{LP}}(\mathbf{90}^{\circ}) \cdot \mathbf{M}_{\mathbf{QWP}}(+\mathbf{45}^{\circ}).\end{aligned} \tag{4.1}$$

Every pixel of the CCD detector records an intensity, which is described by the first component of the Stokes vector [32]. Hence, the physically recorded observable is:

$$I_{\mathrm{L,R}} = [1, 0, 0, 0] \cdot \mathbf{M}_{\mathrm{L,R}} \cdot \begin{bmatrix} s_0 \\ s_1 \\ s_2 \\ s_3 \end{bmatrix}, \tag{4.2}$$

and by inserting the Mueller matrices from Appendix A.1 into Eq. (4.1) it follows that:

$$\begin{aligned} I_{\mathrm{L}} &= s_0 + s_3, \\ I_{\mathrm{R}} &= s_0 - s_3. \end{aligned} \tag{4.3}$$

The CDSI can then be calculated from Eq. (2.31) and it can be shown that the setup is sensitive to the s_3 component of the Stokes vector:

$$\mathrm{CDSI} = \frac{I_{\mathrm{L}} - I_{\mathrm{R}}}{I_{\mathrm{L}} + I_{\mathrm{R}}} = \frac{s_3}{s_0}. \tag{4.4}$$

Here s_0 is the total intensity of the scattered light ($I_{\mathrm{Total}} = I_{\mathrm{L}} + I_{\mathrm{R}}$) and the fraction $\frac{s_3}{s_0}$ equals the *degree of circular polarization* in the Mueller–Stokes formalism [32]. Note that the circular differential scattering intensity is spectrally resolved by the setup: $\mathrm{CDSI} = \mathrm{CDSI}(\lambda)$.

4.2.2 Modeling Linear Polarization Artefacts

For molecular samples [12, 14, 15] as well as artificial nanostructures [10, 11, 18, 21, 22] it is expected that even linear optical anisotropies exhibited by the sample can be falsely detected as a circular differential intensity due to a fixed light-object geometry. In fact, linear polarization effects (LB or LD) can contribute to molecular CD signals as well as the CDSI of plasmonic nanostructures when the observed particle is stationary, because of the excitation and coupling between electric and magnetic dipolar as well as higher order multipolar terms. This means that achiral but optically anisotropic particles, like nanorods, exhibit a chiroptical response which does not originate from chirality intrinsic to the nanoparticle. To qualitatively understand this behaviour a simple calculation based on the Mueller–Stokes formalism was carried out.

Nanorods are known to exhibit linear optical anisotropy, i.e. linear dichroism [24], and therefore a single nanorod attenuates light differently depending if it is polarized along its long or short axis. In general it acts as a linear polarizer but obviously with a poor attenuation ratio between the two orthogonal polarization states. A rotation of the nanorods' long axis with respect to a second linear polarizer, which acts as an analyser, will thus modulate the transmitted intensity dependent on the rods' rotation angle. If the rod is freely diffusing it will permanently change its' orientation with respect to the analyser and a characteristic intensity fluctuation can be measured. This is the basic idea behind the use of plasmonic nanorods as orientation sensors [29], which is e.g. employed in microrheological measurements [25, 26]. It is thus expected that such linear anisotropies have important consequences for the here presented setup with its fixed polarization optics and with the Wollaston prism acting as a linear analyser. In fact, if an achiral but optically anisotropic particle modulates the intensities corresponding to the linear polarization states of the light, the conversion from LP to CP by the quarter-wave plates, will thereby also lead to a modulation of the recorded CDSI signal. Consequently, it is expected that a nonzero CDSI can occur due to linear anisotropies exhibited by an achiral nanorod. Notice that, if a nanorod exhibits LD it will automatically also possess LB, due to the Kramers-Kronig relation.

A chiral nanohelix is associated with circular dichroism (CD) and therefore also circular birefringence (CB). Primarily, it can be thought of as a nanorod which has a small amount of circular anisotropy in addition to its linear polarization properties. It follows that a nanohelix exhibits LD, LB, CD and CB. Hence, it is important to understand how these effects contribute to the signal in the described setup. A general Mueller matrix was introduced as the sample, i.e. a single nanoparticle, which comprises every possible polarization-changing effect (LB, LD, CB, CD). By decomposing it into matrices associated with the individual effects (see also Sect. 2.6) it follows that:

$$\mathbf{M}_{\text{sample}} = \mathbf{M}_{\mathbf{LB}} + \mathbf{M}_{\mathbf{CB}} + \mathbf{M}_{\mathbf{LD}} + \mathbf{M}_{\mathbf{CD}}, \tag{4.5}$$

and the impact onto the CDSI can be modeled separately.

But it's not only the optical anisotropy of the sample itself that plays a role, but rather any optical element of the setup will alter the lights' polarization state and thus contribute to the detected differential intensities. Most prominently, the dark-field condenser is not perfect and will introduce an elliptical polarization to the initially unpolarized light [19, 20]. Clearly, such partially polarized incident light will exacerbate the spurious influences contributing to the CDSI signal. To incorporate the imperfect condenser into the analysis, it was also modeled as a Mueller matrix exhibiting LB, CB, LD and CD effects, equal to $\mathbf{M}_{\text{sample}}$. The reorientation of the nanohelix was considered by assuming that the incident light direction ($\mathbf{k} = k \cdot \hat{\mathbf{z}}$) is perpendicular to the xy-plane in which the nanoparticle was rotating ($\mathbf{k} \perp xy$). Rotation of the helix is modeled such that its long axis is parallel to xy and its direction varies between $\gamma = 0$ and 2π. Consequently, it's orientation changes between orthogonal or parallel alignment with respect to the incident electric field vector $\mathbf{E} \perp \mathbf{k}$. A tilt of the long axis out of the xy-plane would only change the magnitude of the birefringence or dichroism and is therefore omitted in the analysis. The Circular Differential Scattering Intensity in dependence of the angle γ can be deduced by rewriting Eq. (4.1) as:

$$\begin{aligned} \mathbf{M}_{\text{L}} &= \mathbf{M}_{\text{QWP}}(-\mathbf{45}^{\circ}) \cdot \mathbf{M}_{\text{LP}}(\mathbf{0}^{\circ}) \cdot \mathbf{M}_{\text{QWP}}(+\mathbf{45}^{\circ}) \cdot \mathbf{M}_{\text{sample}}(\gamma) \cdot \mathbf{M}_{\text{condenser}}, \\ \mathbf{M}_{\text{R}} &= \mathbf{M}_{\text{QWP}}(-\mathbf{45}^{\circ}) \cdot \mathbf{M}_{\text{LP}}(\mathbf{90}^{\circ}) \cdot \mathbf{M}_{\text{QWP}}(+\mathbf{45}^{\circ}) \cdot \mathbf{M}_{\text{sample}}(\gamma) \cdot \mathbf{M}_{\text{condenser}}. \end{aligned} \tag{4.6}$$

The CDSI was then derived for a constant wavelength $\lambda = \lambda_0$ by inserting Eq. (4.6) into Eqs. (4.2) and (4.4). The results for a rotating particle are visualized in Fig. 4.2 where every column and row displays the sample and condenser effects, respectively.

In this simple approach the anisotropy effects are calculated separately and can lead to a CDSI = 100%. In practice, the situation is more complex and different effects occur simultaneously, maybe cancel or enhance each other. Nevertheless, the influence of technical imperfections are qualitatively illustrated by the calculations. Noteworthy is the 5th column in Fig. 4.2. It displays the conditions of an ideal experiment, i.e. the condenser is non-polarizing and thus only unpolarized light is incident onto the sample. In such a situation only contributions from the sample's CD are relevant for the occurrence of a CDSI and any other anisotropy within the sample leads to CDSI = 0. In case the sample exhibits CD (4th row), a CDSI $\neq 0$ is detected regardless of the incident light's polarization state, as expected. Unsurprisingly, a CD introduced by the condenser (4th column) will also result in nonzero signals irrespective of the optical property of the sample. Besides this, there is only one other combination where the setup detects a nonzero CDSI even if there is no circular dichroism exhibited by either the sample or the condenser. This case arises if the condenser shows LD and the sample LB, which remarkably could be the case for an achiral nanorod. Then, the orientation of the sample additionally causes a modulation of the CDSI, too. Notice, that the reversed combination (condenser LB + sample LD) yields CDSI = 0 because birefringence does not affect the polarization state of the incident light, which is initially unpolarized. Then, the sample's LD would only lead to a linearly polarized state of the scattered light, and hence no circular differential intensity is detected. However, the simple analysis demonstrates that indeed, achiral but optically anisotropic particles can give rise to CDSI signals and that imperfections of the condenser will exacerbate such effects.

It should be emphasized once more that this simple approach using Mueller matrices provides a qualitative analysis but does not capture the full picture of this highly complex system. For example, every other component in the setup (mirrors, grat-

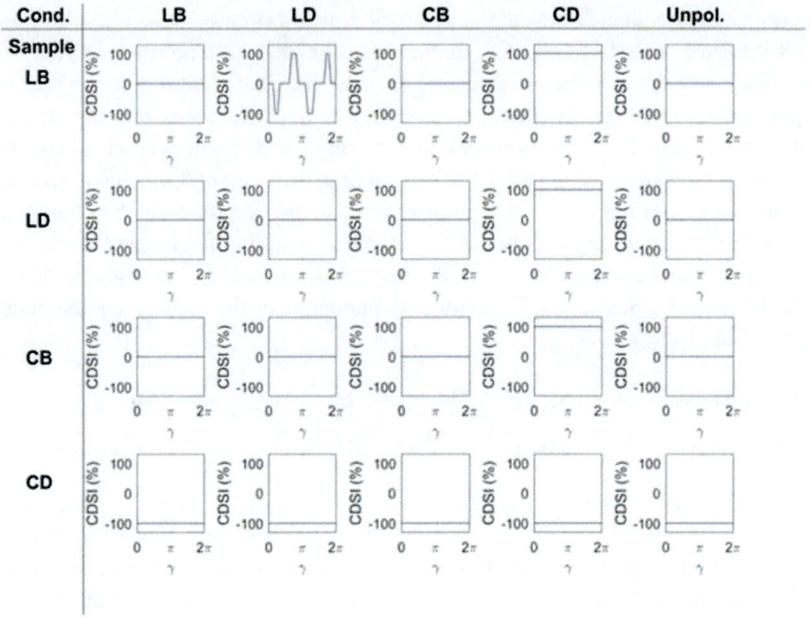

Fig. 4.2 Schematically depicted Mueller–Stokes analysis to shed light on the impact individual optical anisotropies have on the CDSI signals recorded with the dark-field scattering spectroscopy setup (at constant wavelength $\lambda = \lambda_0$). Each entry represents the ratio $\text{CDSI} = s_3/s_0$ under the assumption that s_0 is fully polarized and that a combination of linear and circular birefringence/dichroism effects occur in the condenser (columns) or the sample (rows). The last column shows the ideal case of a non-polarizing condenser, whereas in practice a superposition of mostly unpolarized and slightly polarized light is expected

ing, etc.) will introduce more polarization artefacts. Moreover, in the model it is assumed that the impact of fixed optics will be canceled out by a proper calibration of the instrument (as will be explained below). In practice the dark-field condenser's height had to be adjusted before every measurement and thus it varied its position between experiments and consequently the instruments' calibration changes too. Another simplification is made by assuming that all polarization effects in the sample are independent, e.g. if the nanoparticle reorients with respect to the incident light only the strength of an individual effect is varied. However, in practice a reorientation will likely lead to emergence of other polarization effects, and the decomposition shown in Eq. (4.5) is then invalid. Besides that, the dark-field condenser illuminates the sample with an annular light-cone from 0° to 360° and the incident light has many $\hat{k}$-vectors, possibly carrying different polarization states. However, the qualitative results obtained by the calculations presented here are in full agreement with the experiments shown below and are helpful to understand the observations for differently shaped nanoparticles. Most notably, they underpin the fact that also achiral shaped particles potentially exhibit nonzero CDSI responses owing to linear optical anisotropies.

4.2.3 Spectral Data Processing

In the setup shown in (Fig. 4.1) a single detector acquired intensities corresponding to left- and right circularly polarized light, and the signals have to be processed as is now discussed. A background correction was necessary because not every photon that was collected originated from light scattered off the single particle. On the detector every pixel was accumulating a photon flux, i.e. an intensity I_P, for a fixed exposure time. The intensity being measured on the detector can be approximated as a superposition of several sources:

$$I_P(\lambda) = g(t) \cdot I_{\mathrm{HAL}}(\lambda) \cdot F(\lambda) \cdot [S_{BG}(\lambda) + S_{NP}(\lambda)]\,. \tag{4.7}$$

Here, the single particle was illuminated by a 120 W halogen lamp as light source whose spectrum $I_{\mathrm{HAL}}(\lambda)$ convolved at the detector with the scattering spectrum of the nanoparticle, denoted as $S_{NP}(\lambda)$. The lamp intensity did not only change with wavelength, but also fluctuated over time, which is reflected by the time-dependent function $g(t)$. Moreover, scattering signals from imperfections, e.g. scattered light by out-of focus objects S_{oof}, dust S_d, etc. was also collected. Every scattering source contributed to the measured signal at the detector as a background denoted as $S_{BG}(\lambda) = \sum ... + S_{oof} + S_d$. Every optical element in the optical train also altered the polarization state as the light propagates through the setup. Elements like waveplates, prisms, mirrors, grating etc. are dispersive and the detector itself also exhibited a wavelength-dependent quantum efficiency $QE(\lambda)$. All these factors contributed to the recorded intensity at a specific wavelength, and the various contributions are grouped together as $F(\lambda) = \sum ... + QWP_{-45^\circ}(\lambda) + QE(\lambda)$.

The detector array consisted of 255×1024 pixels in which each row corresponded to the intensity spectrum $I(\lambda)$ of a different area in the sample plane, specified by the pixel size. A pixel is $26 \times 26\,\mu\mathrm{m}^2$ and the $40\times$ objective demagnified it to a size of $650 \times 650\,\mathrm{nm}^2$ in the sample plane. The investigated nanostructures had a characteristic length scale of $\ell \propto 150\,\mathrm{nm}$. It was thus reasonable to assume that only a few rows on the detector are needed to record the spectrum of the particle $I_P(\lambda)$, whereas most of the other rows record background light:

$$I_{BG}(\lambda) = g(t) \cdot I_L(\lambda) \cdot F(\lambda) \cdot S_{BG}(\lambda). \tag{4.8}$$

Those two detector regions were subtracted in order to obtain the pure particle scattering spectrum:

$$\begin{aligned} I_{\mathrm{pure}}(\lambda) &= I_P(\lambda) - I_{BG}(\lambda) \\ &= g(t) \cdot I_L(\lambda) \cdot F(\lambda) \cdot S_{NP}(\lambda). \end{aligned} \tag{4.9}$$

Here it was assumed that the background was constant over the detector rows and the lamp fluctuation over time was small such that $g(t_1) \simeq g(t_2)$. In practice apertures prevented an overlap between LCP and RCP light which is manifested by two spatially separate regions in the raw signal recorded by the detector, which is shown in Fig. 4.3a for a typical experiment. The upper half corresponded to LCP and the bottom half to RCP light. The most intense rows yielded the particle scattering $I_P(\lambda)$, whereas the less intense areas were due to the background $I_{BG}(\lambda)$ and both were also spatially well-separated. The dark rows around the two bright regions corresponded to rows that were blocked by apertures and hence they have been omitted in the further analysis.

During an acquisition the probed nanoparticles were freely suspended and thus randomly varied their position because of Brownian motion. Consequently the intense scattering signal changed its position also on the CCD array between consecutive acquisitions. Therefore it was necessary to first locate the signal in the raw pixel data before proceeding with further evaluation steps. This has been carried out by a Matlab script which automatically detected the most intense row.[1] Then the mean over a constant number of neighbouring rows was calculated in order to extract the particle spectrum $I_P(\lambda)$ (Fig. 4.3b). Only the y-coordinate changed over the course of all frames acquired during one measurement. The most intense rows as well as the unused rows were subsequently subtracted from the raw data to access the background spectrum $I_{BG}(\lambda)$, which was then averaged over a constant number of rows (Fig. 4.3c). Corresponding background-free scattering spectra ($I_{\text{pure}}(\lambda)$) as described by Eq. (4.9) for each of the LCP and RCP areas are shown in Fig. 4.3d. To calibrate the setup, i.e. to correct the measurements for the lamp spectrum and dispersive components, a reference spectrum of a 1.5 μm TiO_2 bead was acquired initially. It is expected to give a constant, spectrally flat scattering spectrum $S_{TiO_2}(\lambda) = c$ and hence:

$$I_{TiO_2}(\lambda) = g(t) \cdot I_L(\lambda) \cdot F(\lambda) \cdot c. \tag{4.10}$$

Dividing the particle spectrum by the reference then yields the de-convolved left- and right-circularly polarized scattering spectrum of a single particle:

$$I_{\text{L,R}}(\lambda) = \left.\frac{I_{\text{pure}}(\lambda)}{I_{TiO_2}(\lambda)}\right|_{\text{L,R}} = \left.\frac{S_{NP}(\lambda)}{c}\right|_{\text{L,R}}. \tag{4.11}$$

The constant factor c canceled out during the following calculation of the CDSI according to Eq. (4.4). Therefore, only light scattered off the particle was contributing to $I_{\text{L,R}}$ in the end. For comparison, the convolved spectra are plotted next to the corrected spectra in Fig. 4.3d, e, respectively. Notice that in order to show several spectra from individual frames in one plot the raw spectra were sometimes smoothed over intervals of adjacent wavelengths (either $\Delta\lambda = 9.13$ nm or $\Delta\lambda = 22.8$ nm). Otherwise the plot would be obscured by noise. Smoothing was only applied for plotting, e.g. in Fig. 4.3d, e, whereas calculations were always carried out using the raw data.

[1] The peak finding was done with the `findpeaks` command.

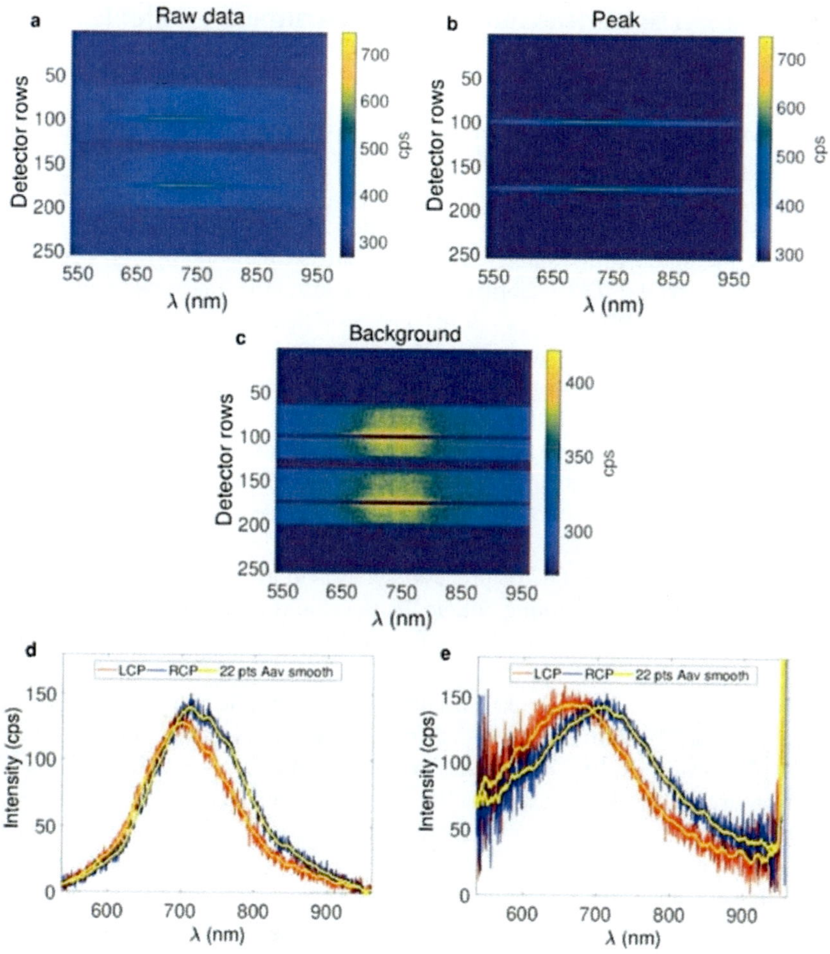

Fig. 4.3 Signal processing of a recorded single particle spectrum. **a** Raw data as acquired by the CCD yields two areas corresponding to LCP and RCP scattered light. **b** Only the most intense rows corresponded to the particle scattering intensity $I_P(\lambda)$ and were averaged. **c** The remaining rows were also averaged and provided the background $I_{BG}(\lambda)$, which was subtracted. **d** The corresponding background-subtracted spectra $I_{\text{pure}}(\lambda)$ were still a superposition, among other things, with the lamp spectrum. **e** Scattering spectra $I_{L,R}(\lambda)$ that were normalized by a whitelight reference (TiO_2) were used to calculate the CDSI. Figure d and e adapted from Ref. [1]

4.3 Comparison with Commercial Instrument (Colloidal Ensembles)

To demonstrate the proper operation of the novel dark-field spectrometer, measurements acquired with this setup were compared to those obtained from a traditional (bulk) CD spectrometer. Spectra of colloidal ensembles consisting of several differently shaped nanoparticles were first recorded with the dark-field setup and subsequently measured in a commercial instrument (*JASCO* J-810). As a reference commercial, achiral (CTAB-stabilized) Au nanorods with a Localized Surface Plasmon Resonance (LSPR) at $\lambda_{LSPR} = 650\,\text{nm}$ (specified by the supplier *Alfa-Aeser*) as well as Au nanospheres with a diameter of $d = 150\,\text{nm}$ (*Sigma-Aldrich*) were chosen. Their particle density was on the order of $n \simeq 10^9\,\text{ml}^{-1}$ and they are expected to show no circular intensity difference. On the other hand, helical shaped Au nanoparticles of opposite handedness served as the chiral sample, i.e. non-zero chiroptical spectra are expected for them. The (2-turn) Au nanohelices were fabricated with GLAD (Sect. 2.7) under a continuous rotation of the substrate around its normal during the deposition process. To facilitate the helix growth Au and Ti have been evaporated 15:1 . After the GLAD process they were immersed into Milli-Q water via ultrasonication and the resulting colloidal suspensions had a concentration around $n \simeq 10^9\,\text{ml}^{-1}$. SEM images of the helical nanostructures after sonication are shown in Fig. 4.4.

For the measurement of (bulk) ensemble spectra with the commercial spectrometer, the colloidal solutions were filled into a $\ell = 1\,\text{cm}$ path length quartz cuvette (*Hellma*) and were probed in transmission, i.e. with the detector mounted at $\theta = 0°$. The measurement was repeated with the detector mounted at $\theta = 90°$ utilizing a 4-side polished quartz cuvette (for the scattering geometry see also Fig. 2.5b). Acquisition of ensemble spectra in the dark-field spectrometer was done in a sample chamber consisting of a sealed coverslip sandwich. 25 μl suspension was pipetted onto a glass coverslip with a double-sided adhesive tape (*Gene Frame*) which bonded to a second coverslip that was pressed from top and thus formed an air-tight seal. For the acquisition of spectra, the circular aperture in front of the first QWP was opened as wide as possible and the slit width was chosen to be 800 μm. This ensured a large

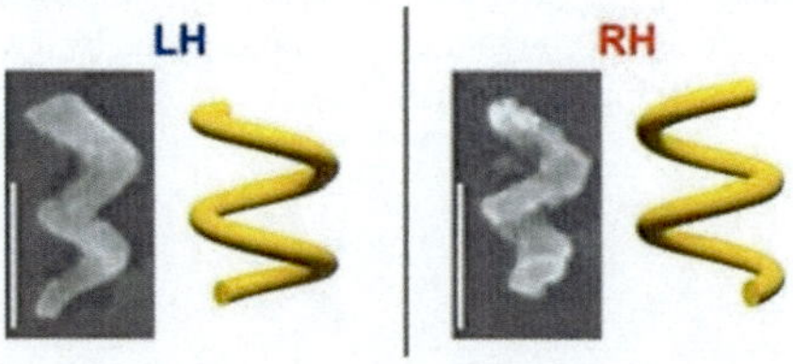

Fig. 4.4 SEM images and schematic visualization of the chiral 2-turn Au nanohelices fabricated by GLAD (see Sect. 2.7) for both, the Right-Handed (RH) and Left-Handed (LH) enantiomer (scale bars 100 nm). Both enantiomers exhibit a waist $d \simeq 70\,\text{nm}$ and height $h \simeq 150\,\text{nm}$. Figure adapted from Ref. [1]

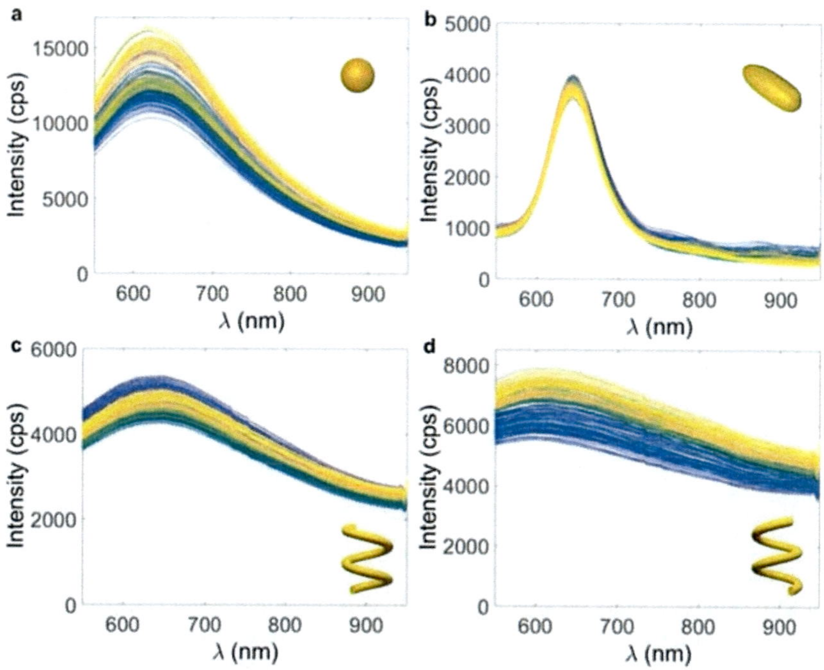

Fig. 4.5 Total scattering intensity ($I_{Total} = I_L + I_R$) spectra of colloidal ensembles of different achiral- and chiral-shaped nanostructures in H_2O that were recorded by the novel dark-field scattering spectroscopy setup. **a** Commercial Au nanospheres ($d = 150$ nm). **b** Commercial Au nanorods ($\lambda_{LSPR} = 650$ nm). **c** LH and d) RH Au nanohelices fabricated by GLAD (corresponding to Fig. 4.4). Figure taken from Ref. [1]

field of view and consequently scattering light from a large area in the sample plane was recorded. 60 rows on the detector contributed to the mean spectra for the LCP and RCP channel, this time without background subtraction. For the 40 × objective that has been used this corresponds to an area of $\sim 40\,\mu$m $\times 20\,\mu$m in the focal plane, and thereby roughly ten particles have been in the field of view during a measurement. The nanostructures' sizes were on the order of $d \sim 150$ nm or smaller. By using Eqs. (2.15) and (2.16) the translational diffusion coefficient for a particle with a hydrodynamic radius $r = 75$ nm in water ($\eta \approx 1$ cP) was calculated to be $D_{trans} \approx 3\,\frac{\mu m^2}{s}$. A CDSI spectrum was acquired in 3 s and 400 spectra were recorded in total. The overall measurement time was thus ~20 min, which was long enough to assume that a large number of different particles diffuse in and out of the field of view during the experiment. Hence, by averaging over the 400 spectra a CDSI spectrum was deduced, that provided an ensemble average similar to the average that is measured in a commercial spectrometer.

Resulting spectra of the total scattering intensity $I_{Total} = I_L + I_R$, which were measured by the dark-field setup, are shown in Fig. 4.5 (note the different y-axes). Achiral Au spheres can be modeled by simple Mie-scattering theory [16] and a broad peak at $\lambda = 650$ nm is expected. Here the resonance is observed at $\lambda = 625$ nm

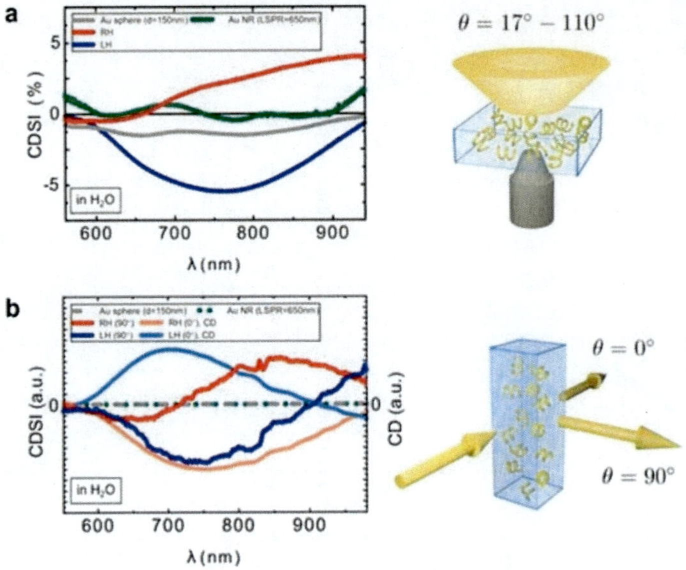

Fig. 4.6 Comparison between chiroptical spectra of colloidal ensembles acquired for differently shaped nanostructures measured in the dark-field microscope as well as a commercial spectrometer. Both instruments measure zero differential intensities for achiral objects (nanospheres and nanorods), whereas the spectra of nanohelices show a peak of opposite sign for the two enantiomers. **a** CDSI accumulated over 400 spectra ($\sim$20 min) measured in the dark-field microscope (corresponding to Fig. 4.5a–d). **b** Measurements in the commercial spectrometer with a scattering angle of either $\theta = 0°$ or $\theta = 90°$ measures CD or CDSI, respectively. Changing the detection angle leads to an expected sign reversal for the helix spectra. Figure adapted from Ref. [1]

(Fig. 4.5a) which suggests that the spheres had a slightly smaller diameter than specified. In case of the achiral nanorods the LSPR at $\lambda = 650$ nm was confirmed (Fig. 4.5b). The chiral Au nanohelices (Fig. 4.5c, d) also exhibited a scattering peak, which was comparable to the spheres in case of the Left-Handed nanoscrews. This is anticipated as its size deduced from the SEM image ($h \simeq 150$ nm) agrees with the diameter of the spheres. The other enantiomer exhibited a slightly broadened scattering resonance, which probably originated from imperfections during the GLAD process, e.g. leading to a polycrystalline structure and surface roughness.

In Fig. 4.6a the CDSI spectrum after accumulating over 400 spectra is depicted. As expected, the achiral samples show zero CDSI within the experimental error of the instrument, whereas the two chiral nanohelices display a peak with opposite sign for the two enantiomers. The LH helix's spectrum is more clearly discerned whereas the RH enantiomers's spectrum peaks at longer wavelengths and appears broadened, which is in accordance with the observations made for the total scattering intensity spectra in Fig. 4.5.

Ensemble measurements conducted with the commercial instrument, for which the accessible wavelength range was larger, are depicted in Fig. 4.6b. As pointed out in Sect. 2.6, commercial instruments are usually employed to measure CD spectra of molecules in transmission, i.e. under a scattering angle of $\theta = 0°$ with respect to the incident light direction. In the case of molecules scattering is negligible and the measurement primarily records absorption. When measured in transmission ($\theta = 0°$), the chiral shapes yield signals of opposite sign, being more pronounced in case of the left-handed helix, whereas both achiral nanostructures exhibit zero CD, as expected. The measurements stem from a very large number of structures and so the samples are spatially isotropic during the acquisition of a spectrum. Again, the signal of the RH enantiomer peaks at longer wavelengths compared to the LH enantiomer.

Any other scattering angle ($\theta \neq 0°$) records scattered light intensities. The measurement at $\theta = 90°$, shown in Fig. 4.6b, had a signal-to-noise ratio that was too low to reliably record a spectrum. However, qualitative comparisons are still possible. The resonance maxima are shifted by roughly $\sim$ 50 nm between a measurement with absorption and scattering configurations and most noteworthy, the sign of the circular intensity difference is reversed. This can be understood by assuming a left-handed sample, which preferentially absorbs and scatters LCP light. Consequently, the recorded intensities at $\theta = 0°$ yield intensities with magnitudes $I_{\mathrm{R}}^{Abs} > I_{\mathrm{L}}^{Abs}$. In contrast, at $\theta = 90°$ more LCP light reaches the detector: $I_{\mathrm{L}}^{Scat} > I_{\mathrm{R}}^{Scat}$. Calculation of the circular intensity differences ($\propto I_{\mathrm{L}} - I_{\mathrm{R}}$) thus exhibits a sign reversal between the measurement of absorbance or scattering. Therefore, the observed sign change for measurements at $\theta = 0°$ and $\theta = 90°$ is also expected for the chiral nanohelices in Fig. 4.6b.

Interestingly, in the dark-field scattering setup the condenser's and objective's NA (1.2-1.4 and 0.9) fixes the scattering geometry to an angle of $\theta = 17°$ and 110° (Fig. 4.6a). In contrast, the scattering based ensemble measurements in the commercial spectrometer have been conducted at $\theta = 90°$ (Fig. 4.6b). This implies that spectra obtained by both instruments are related but not equivalent because the scattering angle differs. However, their results are in good agreement. Spectra of chiral nanoparticles exhibit a peak of opposite sign, and the achiral nanospheres or nanorods don't display circular intensity differences, irrespective of the detection angle or setup. Moreover, the wavelength of the peak maxima of the chiral particles agree qualitatively between both setups. Hence, the spectra measured with the two setups came to the same result and properly determined chirality inherent to the particles' shapes. This clear similarity between the spectra acquired by both the commercial instrument and the newly built dark-field spectroscopy setup thus validated the proper operation of the dark-field for single particle spectroscopy.

4.4 Observation of Single Brownian Nanoparticles

From the preceding discussion it is clear that an oriented sample, including an oriented single nanostructure, will show CDSI signals that are not only due the scatterer's chirality. This leads to problems in the interpretation of chiroptical spectra stemming from a single nanostructure, whose orientation is fixed with respect to the incident electric field vector. Achiral structures can then exhibit non-zero circular intensity differences. To demonstrate this, a freely diffusing single nanoparticle was investigated in solution, away from a surface. Because the particle remained untethered in a liquid, it was free to dynamically change its orientation during the experiment. However, the balanced detection scheme opened up the possibility to observe a spectrum at one specific instant in time during which the particle remained nearly static. Thus, a single spectrum corresponded to a unique orientation of the nanoparticle in 3D. Beyond that, it was possible to acquire a time-series comprising many individual spectra taken at distinct orientations, i.e. observation of one and the same single particle over longer times, as depicted in Fig. 4.7a. A momentarily taken (snapshot) CDSI measurement will then still show linear polarization effects. However, by taking the time-average, $\langle \mathrm{CDSI} \rangle_t$, over a sufficiently long time-series, the particle will statistically sample random orientations and a true chiroptical spectrum is detected and the unwanted linear anisotropies (LB, LD) vanish. Consequently, they play no role for the time-averaged circular differential intensities of achiral objects. Only truly chiral shaped nanostructures will then exhibit a non-zero CDSI, similar

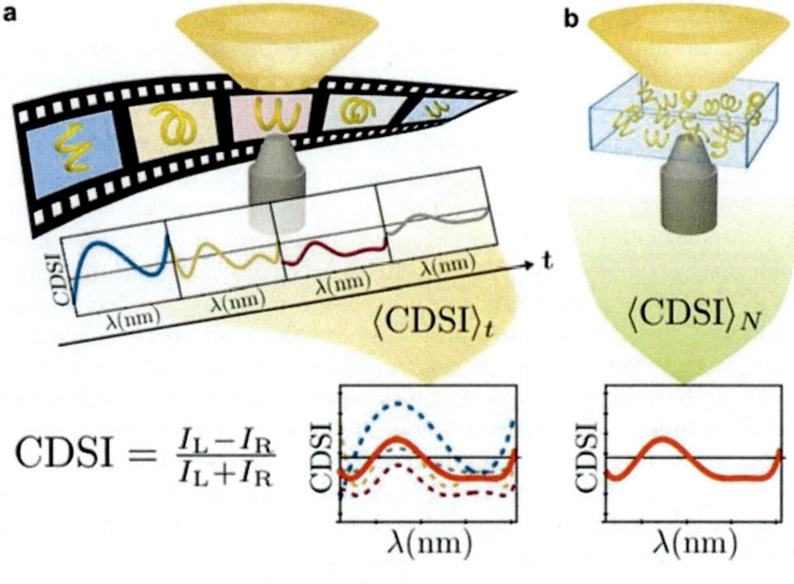

Fig. 4.7 Schematic visualization of single-particle and ensemble chiroptical spectroscopy in the dark-field spectrometer. **a** During the observation of a freely diffusing single nanoparticle the individual snapshots of a time-series experiment exhibit a distorted Circular Differential Scattering Intensity due to a nearly static alignment of the scatterer in one frame. However, the time-average, $\langle \mathrm{CDSI} \rangle_t$, over many observations of the same particle at different orientations yields the true chiroptical spectrum, which unequivocally reflects chirality inherent to the single particle. **b** Schematics of a CDSI measurement taken for a colloidal ensemble containing many identical copies of a particle that are isotropically oriented in space ($\langle \mathrm{CDSI} \rangle_N$). According to the ergodic principle the time-average equals the ensemble-average ($\langle \mathrm{CDSI} \rangle_t = \langle \mathrm{CDSI} \rangle_N$) if particles' orientation is described by the same probability distribution during both measurements. Figure taken from Ref. [1]

to the (bulk) CD measurements on colloidal ensembles. The latter is expected from the ergodic hyothesis.

The ergodic hypothesis states that [33],

> ... for a stationary random process, a large number of observations made on a single system at N arbitrary instants of time have the same statistical properties as observing N arbitrarily chosen systems at the same time from an ensemble of similar systems.

In the context of single nanoparticles this corresponds to the acquisition of one chiroptical spectrum for an individual particle aligned in a specific orientation (Fig. 4.7a). If the particle has equal probability for every direction in 3D space to align, then the average over repeated measurements of this particle will yield the same result as a measurement of an ensemble containing many identical copies of this particle and that are randomly oriented $\langle \mathrm{CDSI} \rangle_t = \langle \mathrm{CDSI} \rangle_N$. The ensemble measurement of N identical particles, $\langle \mathrm{CDSI} \rangle_N$, is schematically shown in Fig. 4.7b. It is important to highlight that ergodicity of a system necessitates an equal probability distribution, here equal probability of the particles' orientation. This is a valid assumption, because in both situations untethered nanostructures are freely suspended in a liquid. Thus they undergo Brownian motion (see Sect. 2.4) leading to a statistically sampled re-orientation. Notice that in practice the observation of an ensemble consisting of identical copies of a nanoparticle cannot be realized experimentally, as unavoidable imperfections during fabrication steps will lead to small deviations of the particles' shapes. Consequently, the time-averaged CDSI spectrum of a single particle in Fig. 4.7a is not exactly the same as observing an ensemble containing a large number of (slightly different shaped) particles. The time-averaged spectrum, $\langle \mathrm{CDSI} \rangle_t$, is therefore a novel observable, which unequivocally deduces the handedness of a single nanoparticle, i.e. chirality intrinsic to the nanoparticle. Remarkably, this observable has never been measured before.

In the experiments of this chapter nanoparticles with sizes of $d \approx 150\,\mathrm{nm}$ have been observed in solution. The exposure time of a spectrum was set to 1 s. Faster exposure times (< 1 s) for spectral acquisitions turned out to be infeasible because the signal was too low. The spectrometer slit width was set to 800 μm, which corresponded to 20 μm in the sample plane. In Sect. 4.3 the translational diffusion constant of particles with $d \approx 150\,\mathrm{nm}$ was estimated to be $D_{\mathrm{trans}} \approx 3\,\frac{\mu\mathrm{m}^2}{\mathrm{s}}$ in water, thus a single nanoparticle would leave the field of view after a few seconds only. Opening the slit even further deteriorated the signal to noise ratio, because too much background light was collected. In order to acquire a time-series over many frames and simultaneously ensure reasonable exposure times for a single frame, it was thus necessary to slow down the nanoparticles' diffusion. This was achieved by immersing the nanoparticles in a 1:20 water-glycerol mixture that had a viscosity $\eta \approx 370\,\mathrm{cP}$ (at 300 K). A particle with diameter $d \approx 150\,\mathrm{nm}$ then exhibited a translational diffusion of $D_{\mathrm{trans}} \approx 8 \cdot 10^{-3}\,\frac{\mu\mathrm{m}^2}{\mathrm{s}}$, which was enough to keep it in focus for several tens of seconds. At the same time the particles' rotational diffusion remained high enough to sample isotropic orientations in space throughout several frames of a time-series experiment, but at the same time the rotation was slow enough to not sample all orientations in space during the acquisition of a single frame.

4.4.1 Chiral Au Nanohelix

Chiral nanohelices were utilized to verify that the time-averaged CDSI provides a measure for a single particles' chirality. To study randomly moving single chiral nanoparticles away from a surface, the colloidal solutions containing Au nanohelices from (Sect. 4.3) were sufficiently diluted. The stock solution was mixed with glycerol (ratio 1:20) to slow down their Brownian motion. From the SEM image in Fig. 4.4 the dimensions were deduced (diameter $d = 70\,\text{nm}$ and height $h = 150\,\text{nm}$) and the corresponding diffusion constants were calculated by approximating the particles as a prolate spheroid and using Eqs. (2.17)–(2.20). Hence, the average translational diffusion constant was $D_{\text{trans}} = 1.25 \cdot 10^{-2}\,\mu\text{m}^2\text{s}^{-1}$, whereas the rotational diffusion times have been calculated as $D_{\text{rot}} = 3\,\text{rad}^2\text{s}^{-1}$ around the short axis and $D_{\text{rot}} = 6.1\,\text{rad}^2\text{s}^{-1}$ around the long axis. The calculations have been confirmed by a numerical calculation done with HYDROSUB, a program that approximates simple shapes with a bead-shell model and then numerically evaluate its diffusion tensors [34]. Thereby $\simeq 2000$ beads modeled a prolate spheroid and reproduce the given diffusion times. If the helix was approximated as a cylinder (with end caps) it follows that the diffusion is even slower ($D_{\text{trans}} = 1.03 \cdot 10^{-2}\,\mu\text{m}^2\text{s}^{-1}$ and $D_{\text{rot}} = 1.55\,\text{rad}^2\text{s}^{-1}$ around the short axis and $D_{\text{rot}} = 3.64\,\text{rad}^2\text{s}^{-1}$ around the long axis). However, both the analytical and numerical diffusion times are only estimates and in practice deviations can occur, e.g. due to local viscosity changes or gravity.

Depending on the model, the rotational diffusion time around the long axis suggests that during one spectrum with one second exposure time half a rotation can occur for this degree of freedom.[2] The helix's rotation around its two short axes only leads to small orientation changes during acquisition of one frame. It is thus expected that re-orientation effects will play a role in the recorded momentary CDSI spectra but that they are generally small enough to not sample all orientations with equal probability. Thus the particle was regarded as approximately stationary during the acquisition of one frame corresponding to one snapshot in Fig. 4.7.

A time-series containing 400 spectra recorded one-after-another was acquired for single Au nanohelices, one for the LH and one for the RH enantiomer. As explained, during the acquisition each of the particles was freely diffusing, which corresponded to a total observation time of more than 6 minutes. The total scattering intensity spectra ($I_{\text{Total}} = I_{\text{L}} + I_{\text{R}}$) are shown for the two enantiomers in Fig. 4.8a, b. Both experiments display distinct resonances that peak around $\lambda \sim 680\,\text{nm}$. Compared to the ensemble measurement shown in Fig. 4.5c the appearance of the resonances are similar but shifted towards longer wavelengths. This is not surprising because the refractive index in water ($n = 1.33$) is lower than for the 1:20 water-glycerol mixture ($n \approx 1.45$) used here. Additionally, now only one nanostructure with a unique shape is contributing and it can exhibit features owing to subtle geometric details that were obscured by the ensemble averaged particle distribution. Thus the single-particle

[2] Note that the mean-squared angular displacement in 3D is $\langle\theta^2\rangle = 4D_{\text{rot}} \cdot \Delta t$ and consequently for $\Delta t = 1\,\text{s}$ one full rotation corresponds to $D_{\text{rot}} = \frac{(2\pi)^2}{4} \simeq 10\,\text{rad}^2\text{s}^{-1}$.

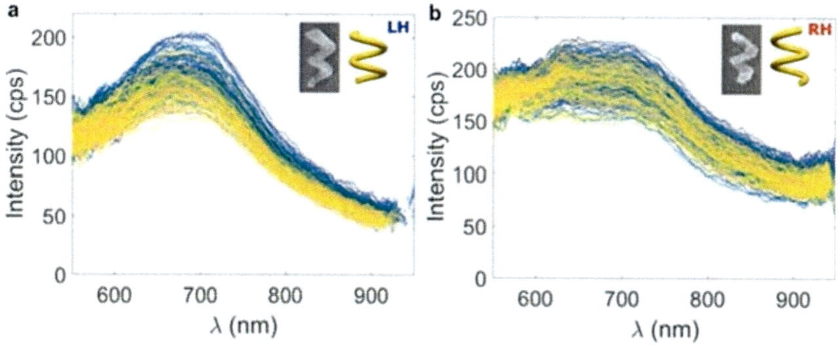

Fig. 4.8 Total scattering intensity ($I_{\mathrm{Total}} = I_{\mathrm{L}} + I_{\mathrm{R}}$) for the left- (**a**) and right-handed (**b**) enantiomer of a single Au nanohelix, measured in the dark-field spectrometer. In both cases the spectra display distinct resonances with maxima around $\lambda \sim 680$ nm. Each line is one individual spectrum (1s exposure) and stays nearly constant throughout the course of the complete experiment, which comprises 400 subsequent frames

scattering spectrum is highly unique and is not necessarily identical with the ensemble spectrum.

Corresponding CDSI spectra for the left-handed nanohelix are shown in Fig. 4.9. Plotting the CDSI at a single wavelength ($\lambda = 750 \pm 5$ nm) over time, here for the first 60 frames, clearly shows that the CDSI keeps changing from frame to frame (Fig. 4.9b). Remarkably, sometimes even a reversed sign is observed between individual frames. In Fig. 4.9b the first 30 full spectra from the same time-series experiment are depicted and they reveal that the fluctuation is not only happening at one single wavelength. Some spectra appear positive overall, whereas on average the CDSI is negative. This finding indicates that each (snapshot) spectrum approximately corre-

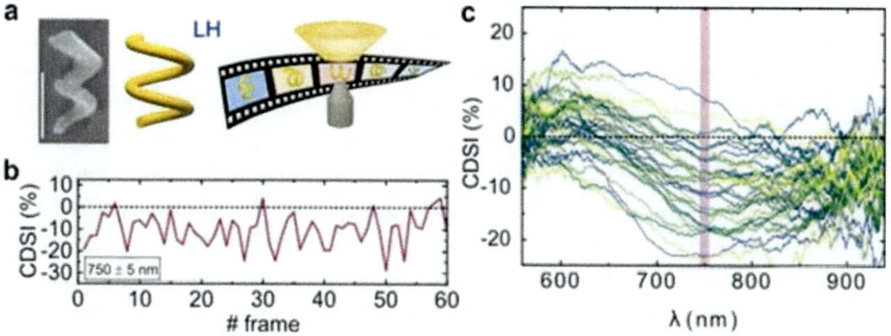

Fig. 4.9 A time-series of CDSI spectra acquired for a randomly re-orienting left-handed Au nanohelix. **a** SEM image and schematics of the time series experiment with a single left-handed nanohelix. **b** CDSI at one wavelength ($\lambda = 750 \pm 5$ nm) plotted for the first 60 frames shows that the intensity difference is fluctuating between individual acquisitions. **c** Corresponding full CDSI spectra (first 30 frames) also vary between frames, and hence indicate a stationary alignment during acquisition of a single spectrum. Figure adapted from Ref. [1]

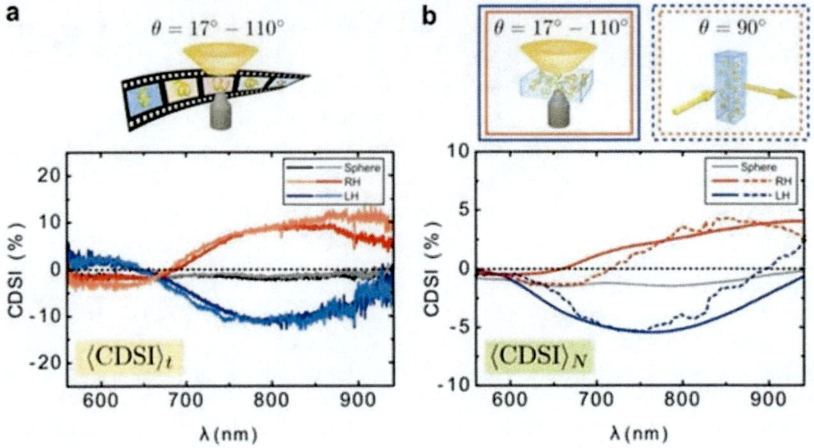

Fig. 4.10 Comparison of single particle time-averaged ($\langle CDSI \rangle_t$) and colloidal ensemble-averaged ($\langle CDSI \rangle_N$) CDSI spectra corresponding to the schematic shown in Fig. 4.7a, b respectively. **a** Time-averaged CDSI for LH and RH Au nanohelices ($t = 400$ frames) as well as Au nanospheres ($t = 60$ frames) display spectra with opposite sign for the chiral enantiomers and no signal for the achiral spheres. **b** Colloidal ensembles of the same nanoparticles observed in the dark-field spectrometer (solid lines) and a traditional CD spectrometer with its detector mounted at $\theta = 90°$ (dashed lines) show similar spectra in agreement with the ergodic principle. Figure adapted from Ref. [1]

sponds to a fixed alignment and thus the above mentioned contributions due to linear polarization effects dramatically impact the form of the CDSI.

This has important consequences as it emphasizes that a spectrum acquired for a fixed orientation does not reflect the true chirality of the nanoparticle. Linear anisotropies can easily distort the spectra and it is thus not possible to deduce the handedness from a single (snapshot) frame with stationary alignment. However, according to the ergodic principle, the time-averaged signal over many frames recovers the same information as the ensemble-averaged bulk CD spectroscopy and hence, it is a true signature of the single particle's chirality. This is verified by Fig. 4.10a which shows the time-averaged mean CDSI, $\langle CDSI \rangle_t$, for Au nanohelices of opposite handedness. Here the average is taken over $t = 400$ frames and corresponds to the data in Figs. 4.8 and 4.9. Additionally, a reference acquired with single achiral nanospheres measured under the same conditions and averaged over $t = 60$ frames is plotted for comparison. As expected, the two chiral enantiomers exhibit CDSI spectra of opposite sign, whereas the CDSI for the achiral spheres is zero. This demonstrates that chirality of a *single* nanoparticle can be unequivocally determined by sampling spatially isotropic observations, i.e. by averaging many instantaneously taken spectra of a diffusing particle over time.

For comparison the ensemble-averaged $\langle CDSI \rangle_N$ of the same samples, dispersed in water, are shown in Fig. 4.10b. The solid lines correspond to the measurement in the dark-field spectrometer, according to Fig. 4.7b, and the dashed lines are the CDSI measured in a cuvette utilizing a commercial spectrometer with it's detector mounted at $\theta = 90°$.[3] Again, the chiral nanoparticles exhibit CDSI spectra with opposite sign for different enantiomers, and the achiral spheres display zero CDSI. Strictly speaking, both ensemble-averaged observations yield different quantities owing to the different scattering angles. However, they are still related and are in qualitative agreement. More important, the spectra of the particle ensembles that were measured in the dark-field spectrometer (solid line in Fig. 4.10b) fully agree with the spectra originating from the single particles' time-averaged CDSI (Fig. 4.10a) measured in the same setup and under the same scattering angles. Only a small shift of the resonance peak is observed due to the change in refractive index of the environment ($n = 1.33$ for water and $n \approx 1.45$ for 1:20 water-glycerol). As already explained, imperfections during nanoparticle fabrication are unavoidable in practice and hence it's impossible to experimentally investigate a pure ensemble of identical nanoparticles. However, the agreement between both experiments clearly indicates the ergodicity of the system: $\langle \mathrm{CDSI} \rangle_t \simeq \langle \mathrm{CDSI} \rangle_N$.

Interestingly, the measurements presented in Fig. 4.10 provide several firsts. On the one hand the first chiroptical spectra of one single nanoparticle that is moving due to Brownian motion and freely suspended in bulk solution, away from a surface, is presented. This is a novel observable that has never been reported previously. On the other hand, the ergodicity of chiroptical spectroscopies was demonstrated, also for the first time. Hence it was verified, that it is possible to deduce the true chiroptical spectrum from a single particle experiment, equivalent to traditional CD-spectroscopy measured on ensembles. Consequently, the time-average ($\langle \mathrm{CDSI} \rangle_t$) recovers the same information as obtained by an ensemble-average ($\langle \mathrm{CDSI} \rangle_N$) over many identical copies. Notice that usage of lower viscosities yield faster rotational diffusion times for the single particle and hence results can potentially be obtained within a few seconds only, for the full spectral range.

4.4.2 Achiral Nanorod and -Sphere

It was just demonstrated that the freely suspended single nanoparticle can provide a true chiroptical spectrum if it is averaged over isotropic orientations. However, it is interesting to further examine the impact of linear polarization effects, which can lead to CDSI signals. They appear for specific orientations and as discussed above (see Sect. 4.4.2) imperfections in the setup will exacerbate such influences. Linear optical anisotropy, e.g. found in nanorods, is enough to momentarily display false and non-zero CDSI spectra if they are measured stationary with a fixed alignment in space. In order to test this hypothesis, experiments with Au nanorods and Au nanospheres have been conducted under the same conditions as for the nanohelices, i.e. they were freely suspended in liquid and thus underwent random Brownian motion. Notice that spheres are obviously achiral, but importantly they do not exhibit

[3] Data acquisition of ensemble-averaged measurements were explained in detail in Sect. 4.3 and the identical data was already shown in Fig. 4.6.

optical anisotropy and hence their rotational diffusion should not lead to any circular intensity difference.

First, Au nanorods were fabricated using the GLAD (Sect. 2.7) technique without rotation of the substrate during evaporation. A SEM image is shown in Fig. 4.11a together with a schematic drawing and 60 total scattering intensity spectra (I_{Total}) of a time-series experiment according to Fig. 4.7. The plasmon resonances of Au nanorods are closely related to the geometric dimensions and shape. For the particle shown, the maximum was found at $\lambda = 800\,\mathrm{nm}$. From the microscope images the height and diameter was, respectively, extracted as $h = 110\,\mathrm{nm}$ and $d = 30\,\mathrm{nm}$. Again, approximating the particle as a prolate spheroid (Eqs. (2.17)–(2.20)) the average characteristic diffusion constant in 1:20 water-glycerol ($\eta \approx 370\,\mathrm{cP}$) was calculated as $D_{\mathrm{trans}} = 2.21 \cdot 10^{-2}\,\mu\mathrm{m}^2\mathrm{s}^{-1}$. The rotational diffusion time around the short axis is $D_{\mathrm{rot}} = 11.9\,\mathrm{rad}^2\mathrm{s}^{-1}$, whereas the rotation around the long axis is $D_{\mathrm{rot}} = 49.4\,\mathrm{rad}^2\mathrm{s}^{-1}$. Again, HYDROSUB [34] confirmed the values and provided slower estimates for the diffusion times in case of a cylinder ($D_{\mathrm{trans}} = 1.86 \cdot 10^{-2}\,\mu\mathrm{m}^2\mathrm{s}^{-1}$ and $D_{\mathrm{rot}} = 6.94\,\mathrm{rad}^2\mathrm{s}^{-1}$ around the short axis and $D_{\mathrm{rot}} = 30.21\,\mathrm{rad}^2\mathrm{s}^{-1}$ around the long axis). This sufficiently suppresses the translational motion, while the rotational diffusion was still relatively fast.

Due to symmetry a rotation around the rod's long axis is not expected to change the light-object geometry and the corresponding spectra. Conversely, a rotation around the short axis should have an impact during one single spectrum, which was acquired with an exposure time of one second. Depending on the model, the rod on average exhibits roughly one full rotation around this degree of freedom. However, it is not expected that all orientations are always sampled equally within one second and thus the CDSI of one snapshot generally comprises a non-isotropic average of orientations. This is in agreement with Fig. 4.11b which shows the CDSI at the resonance maximum ($\lambda = 800 \pm 5\,\mathrm{nm}$) for the first 60 frames of a time-series. Only at some instants in time the CDSI is zero and similarly to the chiral samples before, the signal fluctuates, changes sign and exhibits substantial CDSI signals for some frames. Remarkably, the corresponding full spectra (Fig. 4.11c) reveal comparable magnitudes of $\pm 20\%$ to the CDSI of chiral nanohelices that were shown in Fig. 4.9c. However, if the CDSI gets averaged over a large number of observations, i.e. over spatially isotropic orientations, the signal vanishes to zero as seen in Fig. 4.11d where the time-average, $\langle \mathrm{CDSI} \rangle_t$, is plotted over $t = 60$ subsequent spectra. This is in agreement with the ensemble-average in water, which was again observed in the dark-field spectrometer as well as in a traditional cuvette experiment in transmission[4] (Fig. 4.11e). As expected, all the measurements came to the same result that an achiral nanoparticle does not exhibit circular intensity differences if an average from a time-series of chiroptical single particle spectra is measured.

The impact of linear birefringence and dichroism on the occurrence of a misleading CDSI is evident if the single nanorod is compared with a single nanosphere.

[4] Note that if the CD, measured at $\theta = 0°$ is zero, it follows that the CDSI at $\theta \neq 0°$ is zero too, and it was thus not necessary to repeat the traditional cuvette experiment at a scattering angle of $\theta = 90°$.

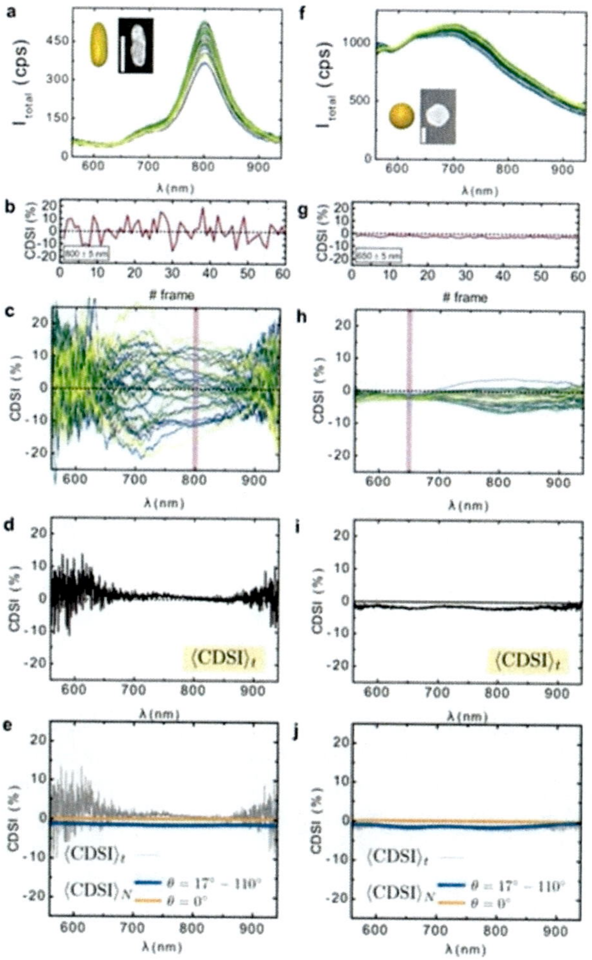

Fig. 4.11 Time-series measurements of CDSI spectra of achiral shaped Au nanorods and nanospheres that randmomly move due to Brownian motion. **a** and **f** Total scattering intensity and SEM together with schematics for the nanorod fabricated via GLAD and a chemically synthesized nanosphere. **b** and **g** CDSI at a single wavelength equal to the plasmon resonance ($\lambda_{rod} = 800 \pm 5$ nm, $\lambda_{sphere} = 700 \pm 5$ nm) for the 60 first subsequent frames of the sequence. **c** and **h** Corresponding full spectra confirm the appearance of non-zero (rod) and zero (sphere) CDSI during individual frames. **d** and **i** Time-average $\langle \text{CDSI} \rangle_t$ over the $t = 60$ subsequent spectra reveal a vanishing CDSI for both achiral shapes. **e** and **j** Measurements conducted on colloidal suspensions in the dark-field spectrometer as well as a cuvette ($\theta = 0°$) reveal the ensemble-averaged $\langle \text{CDSI} \rangle_N$ and come to the same result of zero CDSI, as expected. Figure adapted from Ref. [1]

Therefore the same experiment as just described was repeated with the commercially bought and chemically synthesized nanospheres of diameter $d = 150\,\text{nm}$ mentioned above (see Sect. 4.3). The translational diffusion of nanospheres in water-glycerol (1:20) was $D_{\text{trans}} = 7.9 \cdot 10^{-3}\,\mu\text{m}^2\text{s}^{-1}$, according to Eq. (2.16) and the results of the time-series of spectra are presented in Fig. 4.11f–j. Both particles are geometrically achiral, but in contrast to the rod, the sphere's optical properties are isotropic and hence it neither exhibits LB nor LD. The total scattering intensity peaks around $\lambda = 700\,\text{nm}$ (Fig. 4.11f), which is shifted in comparison to the ensemble measurement in Fig. 4.5a. Again, this is expected by theory [16] owing to the change in refractive index of the surrounding medium here ($n = 1.45$). Accordingly, full CDSI spectra show negligible fluctuations ($< \pm 5\%$) also for instantaneous snapshot measurements, irrespective whether a single wavelength (Fig. 4.11g) or a full spectrum (Fig. 4.11h) is observed. Consequently, the time-averaged CDSI over 60 spectra (Fig. 4.11i) as well as the ensemble-averaged CDSI (Fig. 4.11j) is zero, too. Notice, that the measurement of a spherical nanoparticle validated once more the proper functioning of the novel dark-field setup by providing a reference for zero CDSI. It also proves that the detected chiroptical response of a single-particle reflects its intrinsic chirality.

4.4.2.1 Chemically Synthesized Au Nanord

There is one more interesting feature the nanorods' CDSI displayed in Fig. 4.11. Outside of its plasmon resonance ($\lambda = 800\,\text{nm}$), the recorded scattering signals are very weak and virtually only noise is recorded by the CCD detector. Due to their polycrystalline structure and rough surface, the nanorods fabricated by GLAD exhibited a broadened plasmon resonance compared to chemically synthesized nanostructures. The latter usually show distinct spectra with narrow peak widths, limiting the useful spectral range. However, because many other chiroptical experiments on single-particles that are found in literature [19, 20, 35–37] utilize chemically synthesized nanorods, it was interesting to investigate such samples also in the novel dark-field setup here.

The measurements of the previous section have therefore also been conducted with commercial Au nanorods that have been chemically synthesized. The same particles have already been used for comparison of the dark-field spectrometer with the commercial instrument (see Sect. 4.3). The supplier specified the rods diameter ($d = 22\,\text{nm}$) and height ($h = 55\,\text{nm}$) as well as the LSPR to be $\lambda_{\text{LSPR}} = 650\,\text{nm}$. From the geometric dimensions the diffusion constants were calculated and yield $D_{\text{trans}} = 3.7 \cdot 10^{-2}\,\mu\text{m}^2\text{s}^{-1}$, $D_{\text{rot}} = 70.7\,\text{rad}^2\text{s}^{-1}$ around the short axis and $D_{\text{rot}} = 173.6\,\text{rad}^2\text{s}^{-1}$ around the long axis, again under the assumption of a prolate spheroid (Eqs. (2.17)–(2.20)) in a 1:20 water-glycerol mixture. This time the cylinder model of `HYDROSUB` estimated a translation of $D_{\text{trans}} = 3.06 \cdot 10^{-2}\,\mu\text{m}^2\text{s}^{-1}$, a rotational diffusion around the short axis $D_{\text{rot}} = 37.6\,\text{rad}^2\text{s}^{-1}$ and around the long axis $D_{\text{rot}} = 104.2\,\text{rad}^2\text{s}^{-1}$. Two single nanorods that undergo free Brownian motion were then once again spectroscopically observed with 1s exposure time for a single frame.

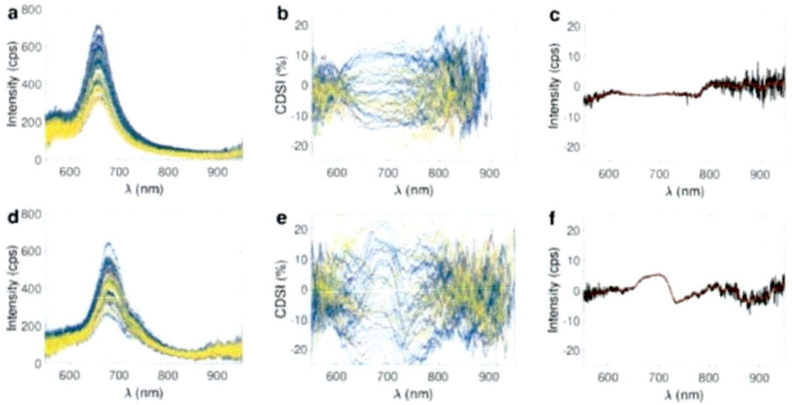

Fig. 4.12 Observation of CDSI spectra of chemically synthesized Au nanorods. **a** and **d** 40 total scattering intensity spectra of chemically synthesized Au nanorod with an LSPR at $\lambda_{LSPR} =$ 650 nm. **b** and **e** Around the resonance the CDSI of individual frames during a time-series yield substantial signals, whereas beyond the resonance virtually only noise contributes (40 frames plotted and data with Background > Peak was omitted). **c** and **f** Averaging the CDSI over $t = 400$ spectra recovers the true chiral spectrum, $\langle \mathrm{CDSI} \rangle_t$, which is zero as expected. Figure **a** to **c** adapted from Ref. [1]

Every 10th frame of a time-series with $t = 400$ frames in total is plotted in Fig. 4.12. The total scattering intensity (Fig. 4.12a, d) and the corresponding full CDSI spectra (Fig. 4.12b, e) for both rods are in agreement with the results obtained for particles prepared by GLAD (see Fig. 4.11). The plasmon resonance is indeed narrower and found at $\lambda \approx 650$ nm, whereas the CDSI momentarily displays non-zero signals but averages to zero over time within the accuracy of the experiment (Fig. 4.12c, f). Interestingly, the rotational diffusion around the short axis suggests that the nanorod exhibits several full rotations within one second and hence a rotationally averaged spectrum is anticipated even for one frame. However, the experiment indicates that during an exposure time of one second not all orientations were sampled equally and thus again a non-isotropic spectrum is observed.

Both particles display distinct spectral features and show minute deviations which are obscured in an ensemble-average, highlighting once more the unique character of the single-particle chiroptical time-series experiment. For the second particle (Fig. 4.12d–f) a small shoulder is visible in the total scattering intensity at $\lambda = 750$ nm, which does not average out and which gives a robust CDSI spectral peak. This suggests that even very small variations in the shape (or due to the environment) can introduce an anisotropy that can be observed in chiroptical single particle spectroscopy.

Importantly it is seen that momentarily taken spectra as well as the time-averaged CDSI only provide reliable data for those wavelength ranges where enough photons were detected. The anticipated smaller useful wavelength range, here $\sim 620 - 720$ nm versus $\sim 650 - 850$ nm for the GLAD rods, is apparent when comparing Figs. 4.12 to 4.11. Beyond this interval it is clearly not meaningful to interpret the data because only noise contributes. This also implies that it is a key prerequisite to normalize the differential circular intensities ($I_L - I_R$) by the total scattering intensity ($I_L + I_R$) in order to allow the comparison between the spectra from different particles and in order to ensure a proper interpretation. Sometimes this is ignored by other studies that either do not normalize their data at all or use other normalization procedures. Examples of such questionable practices include that, for instance the data has been scaled to lie between 0 and 1 or the differential intensity has been divided by the maximum intensity of one channel only $max\{I_L, I_R\}$. In particular the latter is highly questionable because it can artificially cancel out noise beyond the plasmon resonance. Consequently the CDSI is apparently zero although in practice no photons are detected and hence no meaningful interpretation is allowed. This highlights that not only data acquisition but also its interpretation can lead to incorrect chiroptical spectra.

4.5 Single Nanohelix with External Orientation Control

The time-series experiments in the previous sections have been performed on freely suspended nanostructures and under the assumption that their random Brownian motion induces a statistical sampling of arbitrary orientations. To check this, and to confirm the orientation-dependence of the single-particle CDSI, experiments have been performed with particles whose orientation was controlled externally, while they were still freely suspended in bulk solution and away from a surface. For this, magneto-plasmonic helices were prepared and measured whilst their orientation was aligned by means of a weak external magnetic field. The experiments presented in this section were conducted on a right-handed nanohelix that had a small magnetic section at one of its ends. The magnetic part comprised a Ni:Cu pillar (atom ratio 10:1) that was fabricated using GLAD with a fast substrate rotation. On top, the same 2-turn Au:Ti (15:1) helix from previous experiments has been deposited. The resulting geometry ensured that the magnetic moment **m** lies along the longer dimension of the Nickel pillar, and thus along the helix's long axis. A schematic illustration together with a SEM image is shown in Fig. 4.13a. Additionally, the schematics visualizes the external magnetic field vector **B** that exerts a torque on the magnetic dipole, which aligns the nanohelix's long axis parallel to the external field vector. The external field had a strength of 0.38 mT and was generated with the same three-axis Helmholtz-coil setup used for the rotation-translation coupling experiments (Sect. 3.5), providing the possibility to force the particle onto various orientation sequences in 3D space. More details about the coil setup are described in Appendix A.2. However, although the particles' orientation was controlled externally, it was again necessary to suppress its translational Brownian motion, thus it has been immersed in 1:20 water-glycerol prior to the experiment.

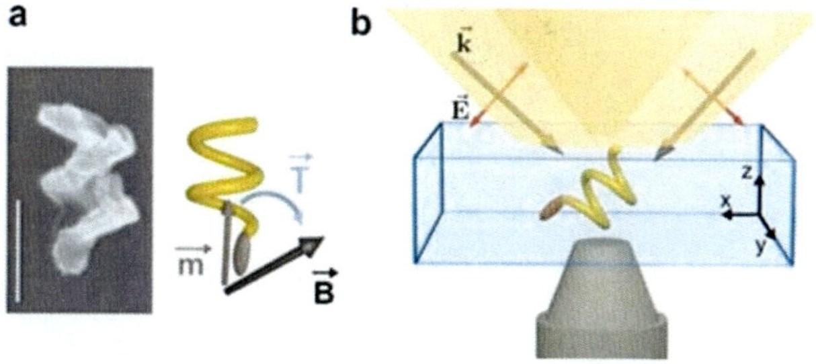

Fig. 4.13 Magneto-plasmonic nanohelix that is controlled by means of a weak external field. **a** SEM image (scale bar 100 nm) and schematic drawing of a (right-handed) Au nanohelix, which has a magnetic segment on one end. In an external magnetic field a torque $\mathbf{T} = \mathbf{m} \times \mathbf{B}$ is exerted on the magnetic moment **m** which will thus align the helix's long axis parallel to the magnetic field vector **B**. **b** Schematic visualization of the light-object geometry, which exhibits a 360° rotation symmetry around the z-axis and yields electric field vectors **E** that are incident under a fixed angle with respect to the xy-plane. Figure taken from Ref. [1]

4.5.1 Rotation in 2D Plane

First, a single magneto-plasmonic nanostructure was rotated in one plane, i.e. in 2D. The magnetic field vector moved on a circular trajectory with fixed rotation frequency $f_{\text{ext}} = 0.1$ Hz. Simultaneously, 100 subsequent CDSI spectra were recorded with an exposure time of 1s. The first 30 spectra from the time-series with the field rotating in xz are shown in Fig. 4.14b, whereas Fig. 4.14c shows the spectra of the same particle for a rotation in the xy-plane. By comparing both plots it can be seen that the CDSI spectra exhibit fluctuations over time in both experiments, in agreement with previous findings. However, a difference between the spectra taken along the two different rotation-planes is apparent, too. Similarly their time-averaged mean $\langle \text{CDSI} \rangle_t$ over $t = 100$ frames differs considerably. In the same manner as before, the CDSI at a single wavelength $\lambda = 800 \pm 5$ nm was analysed. This time a Fast Fourier Transform (FFT) was calculated for the time-series, revealing the frequency domain of the observed fluctuations at this wavelength, which is shown in Fig. 4.14c. Obviously, two distinct peaks at $f = 0.1$ Hz and $f = 0.2$ Hz are observed for the rotation in xz, whereas no peak occurs for a rotation in xy. This indicates characteristic changes in the CDSI with twice the driving frequency, $2f_{\text{ext}} = 0.2$ Hz, and $f_{\text{ext}} = 0.1$ Hz in the former, and only random variations of the spectra in the latter case. The findings can be explained by the geometry of the annular dark-field excitation and the symmetry of the nanohelix as follows.

The conical illumination geometry of the condenser (see Fig. 4.1b) is rotationally symmetric around the z-axis and the incident lights' electric field vectors **E** have a fixed inclination angle with the xy-plane. The helix's alignment with respect to the electric field vectors **E** is thus continuously varying during a rotation if the long helix axis lies in the xz-plane. Therefore the scattering-geometry, i.e. the incident and scattering angle with respect to the helix's long axis, also changes for subse-

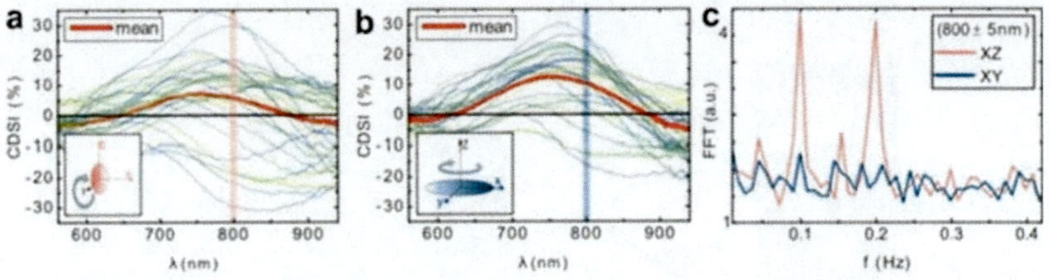

Fig. 4.14 Observation of CDSI spectra of a single Au nanohelix containing a small magnetic moment that was forced on a circular trajectory by an external rotating magnetic field with frequency $f_{ext} = 0.1$ Hz. **a** and **b** By rotating the external magnetic field in the plane parallel to xz (**a**) or xy (**b**), distinct fluctuations are observed for individual spectra (randomly selected) as well as different mean values $\langle \text{CDSI} \rangle_t$ (over $t = 100$ frames). **c** Corresponding FFT analysis at $\lambda = 800 \pm 5$ nm reveals either two characteristic frequencies ($2f_{ext} = 0.2$ Hz, $f_{ext} = 0.1$ Hz) or random noise only. Figure taken from Ref. [1]

quent frames of the time-series resulting in different CDSI spectra. A finite 2-turn helix is formally C_2-symmetric but here the magnetic section on one end as well as imperfections during fabrication renders it C_1 in practice. Therefore, the CDSI is sensitive to alignments orthogonal or parallel to the xy-plane, depending whether the magnetic segment points up- or downwards. This explains why there are two characteristic peaks at $2f_{ext}$ and f_{ext} (Fig. 4.14c). On the contrary, if the helix is rotating parallel to the xy-plane, the light-object geometry remains identical during the rotation sequence. The continuous re-alignment does not change the scattering spectra because of the condensers' rotation-symmetry. Thus, such rotation is indistinguishable in CDSI spectra and the fluctuations occurring are only due to statistical nature, e.g. if the helix' long axis is tilted out of the xy-plane due to random thermal noise. Consequently, no characteristic frequency is anticipated and observed. It follows that the FFT analysis of both measurements is in agreement with the expectations for a nanohelix, which turns in-sync with the rotating external magnetic field. This indicated that the external field can indeed be used to force the magneto-plasmonic nanohelix to undergo a pre-defined re-orientation sequence, even in the highly viscous water-glycerol mixture.

4.5.2 Spatially Isotropic Sampling in 3D

After verification of the external orientation control, a time-series of CDSI spectra was recorded whilst the same nanoparticle was oriented isotropically in 3D-space. This was done by moving the external field along a spherical trajectory that consisted of $i = 100$ points, which were distributed evenly on a unit sphere. Adjacent points were connected, i.e. two points with minimal distance in between, to guarantee a smooth trajectory without discontinuities as schematically depicted in Fig. 4.15a.

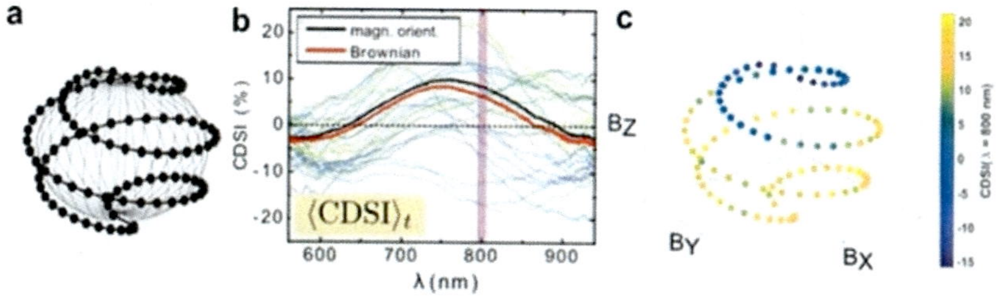

Fig. 4.15 Chiroptical spectroscopy utilizing an externally induced isotropic sampling of orientations of a single magneto-plasmonic nanohelix. **a** By evenly distributing i points on a unit sphere and connecting adjacent points, a spherical trajectory that samples isotropic orientations in space was realized. **b** 30 randomly selected CDSI spectra from a time-series of 10 repetitions of a spherical trajectory with $i = 100$ points. The time-average of the forced isotropic sampling ($\langle \mathrm{CDSI} \rangle_t$, $t = 1000$ frames) is in good agreement with the time-average of the same particle undergoing Brownian motion ($\langle \mathrm{CDSI} \rangle_t$, $t = 400$ frames). **c** Corresponding CDSI at $\lambda = 800 \pm 5$ nm mapped onto the external field coordinates $\{x, y, z\}$ reveal the strong orientation-dependence of chiroptical spectroscopy (average of 10 spectra per point). Figure taken from Ref. [1]

Because of the viscous environment and the low torque, the helices might not align instantaneously with the external field. To ensure the particle was aligned properly during the acquisition of spectra, the external field vector was not varied continuously over the course of the experiment. Instead, it has been moved to one specific orientation with coordinates $\{x, y, z\}$ and after a short waiting time a spectrum was acquired, again with 1s exposure time, before the field moved on to the next point. Synchronization between the spectrometer and the coil setup was realized by a Transistor-Transistor-Logic (TTL) pulse and more details about synchronization and the spherical trajectory are given in Appendix A.2.1. Consequently, during one spherical trajectory sequence the helix's orientation was forced to sample 100 different orientations, with an isotropic distribution. The sequence was repeated 10 times, leading to a time-series comprising 1000 frames in total, which corresponds to an observation time of one particle for more than 16 min. The time-averaged CDSI as well as 30 randomly selected spectra from the time-series are shown in Fig. 4.15b. Clearly, individual frames exhibit distorted CDSI spectra and the time-average $\langle \mathrm{CDSI} \rangle_t$ ($t = 1000$ frames) is positive overall, as expected for a RH helix. Subsequently the external magnetic field was switched off and a time-series of CDSI spectra for the same particle undergoing Brownian motion was acquired, whose time-average $\langle \mathrm{CDSI} \rangle_t$ (now with $t = 1000$ frames) is plotted for comparison in Fig. 4.15b. Both mean values are in excellent agreement and indicate that the spherical trajectory pattern indeed aligns the helix isotropically in space.

Measuring the CDSI with an externally controlled alignment yielded the possibility to map individual spectra onto the orientation of the external magnetic field vector and see how $\{x, y, z\}$ coordinates correlate with them. Accordingly, Fig. 4.15c shows the CDSI at $\lambda = 800 \pm 5$ nm assigned to the field coordinates. Every measurement point is an average over 10 spectra, which were acquired at the same coordinate but with a temporal distance of 100 s in between due to the external reorientation sequence. Strength and sign of the CDSI can be deduced from the color and hence the orientation-dependence of the CDSI is apparent with a tendency to be maximal at the poles of the sphere. The polarity changes roughly around the equator which is in agreement with the symmetry of the annular dark-field condenser (see arguments discussed above). If the helix's long axis changes in the xy-plane no spectral change is expected, whereas a tilt out of this plane, towards the poles, introduces a change in the CDSI. Additionally, the spectral observation is sensitive to whether the magnetic handle points up or down. Consequently an alignment near to the equatorial plane yields smaller differences in the CDSI, and the largest changes are expected at the poles.

The observations for single nanohelices with external orientation control are therefore in full agreement with the findings on Brownian nanoparticles presented in the previous sections. The results highlight once again that only isotropic sampling of orientations leads to a true chiroptical spectrum with vanishing contributions due to linear polarization effects. Consequently, measurement of stationary objects can by no means provide an observable equivalent to a bulk ensemble spectrum.

4.6 Conclusions and Outlook

Chiroptical signals can in general occur due to molecular transitions or, in nanostructures, due to plasmonic coupling effects. The former is subject to classical ensemble CD spectroscopy, established for decades, and typically conducted on an ensemble in a cuvette. It is thus simple to interpret, whereas the measurement of nanostructures introduces considerable complications, especially if single particles are observed on a surface. However, with one exception all previous studies on this topic rely on a circular differential scattering measured on stationary single particles, introducing a fixed light-object geometry. Consequently, their chiroptical responses do not only originate from chirality intrinsic to the nanoparticle, but rather linear polarization artefacts play an important role, i.e. such measurements are by no means equivalent to classical ensemble CD responses. This chapter reports the first observation of a chiroptical response from a single nanoparticle. It is shown that the measured spectra are equivalent to a classical ensemble CD spectrum. In fact, the first true chiroptical spectrum of a single-particle, acquired over spatially isotropic orientations of a single nanoparticle in bulk solution and away from a surface was measured.

A new spectrometer, based on dark-field microscopy and a balanced detection scheme was built and permitted the simultaneous measurement of scattering intensities corresponding to LCP and RCP light. In order to validate the proper functioning of this approach the device was first compared to a commercial instrument by measuring colloidal suspensions of nanoparticles. Further, a time-series of spectra corresponding to individual particles that underwent Brownian motion was measured, thereby inducing a statistical sampling of their orientation over spatially isotropic alignments. The average over such time-series ($\langle \text{CDSI} \rangle_t$) has been demonstrated to equal the results obtained by measuring ensembles ($\langle \text{CDSI} \rangle_N$), as now the artefacts due to linear birefringence and linear dichroism vanish. Thereby ergodicity of chiral spectroscopies has been demonstrated for the first time, implying that the single-particle approach yields the same information as traditional chiroptical spectroscopy measured on ensembles.

Beyond chiral nanohelices for which CDSI spectra are expected, also achiral but optically anisotropic nanorods have been investigated. For the rods it was shown that they can momentarily exhibit large CDSI signals for a measurement at one specific instant in time, which corresponded to a specific alignment in space. In fact, this is equal to measuring a stationary particle and thus contributions of linear polarizations impact the spectra. However, thereafter it was proven that for achiral particles the time-averaged CDSI vanishes. Experiments performed with optically isotropic nanospheres showed that there is no CDSI at all, because now linear effects play no role. The crucial orientation-dependency of CDSI spectra even for chiral objects was further analysed by investigating a magneto-plasmonic nanohelix whose orientation was controlled externally by means of a magnetic field. It was once again verified that only a statistical sample of an isotropic distribution of alignments will provide a spectrum that is a true reflection of the single particles' chirality, in agreement with the ergodic principle. However, again individual spectra can significantly differ from the mean due to linear polarization effects (LB and LD).

Compared to previous reports in literature the presented approach provides a much simpler system for chiroptical spectroscopy without the requirement of complicated correction schemes or any a priori knowledge of the light-object geometry. Remarkably, the approach does not use a large collection of particles or a large volume to unequivocally determine the handedness of the object as is the case for classical CD spectroscopy. Also, in this thesis the CDSI was recorded with the nanoparticles immersed in a highly viscous medium, but lower viscosities yield much faster rotational diffusion times and hence it is potentially possible to deduce a single particles' chirality (and spectrum) only in a few seconds.

The presented experiments only concerned the orientation of the nanoparticles' long axis, i.e. the pitch and yaw degree of freedom. Obviously, a nanorod has C_∞ rotation symmetry around its long axis. Conversely, a helix only has C_1 symmetry around the long axis and hence it should also exhibit different chiral spectra when changing it's rotation around this axis. It is thus highly desirable to undertake more experiments that examine the influence of the nanohelix's roll degree of freedom onto the measured CDSI spectra. Moreover, chiroptical spectroscopy of single nanoparticles is envisioned to be used for the detection of biomolecules, for which it is prerequisite to understand the contribution of plasmonic effects onto the acquired spectra. Recently, the detection of biomolecules with the help of single plasmonic nanorod aggregates was reported [35]. However, again static particles immobilized on a surface have been measured to shed light onto the observed dichroism and whether it arises due to plasmonics (electric dipole-dipole coupling) or scattering-detected circular dichroism of molecules (analogous to fluorescence-detected CD). The latter has long been proposed and theoretically described for ensembles rather than single particles [38]. The straightforward approach presented here is a promising platform for chiral spectroscopy and to observe the influence of chiral molecular analytes *via* individual plasmonic nanostructures.

References

1. Sachs J, Günther J-P, Mark AG, Fischer P (2020) Chiroptical spectroscopy of a freely diffusing single nanoparticle. Nat Commun 11:4513
2. Berova N, Polavarapu PL, Nakanishi K, Woody RW (2012) Comprehensive chiroptical spectroscopy: applications in stereochemical analysis of synthetic compounds, natural products, and biomolecules, vol 2. Wiley, Hoboken
3. Beychok S (1966) Circular dichroism of biological macromolecules. Science 154(3754):1288–1299
4. Grosjean M, Legrand M (1960) Appareil de mesure du dichroisme circulaire dans le visible et lultraviolet. Comptes rendus Hebdomadaires des Seances de L'Academie Des Sciences 251(20):2150–2152
5. Velluz L, Legrand M, Tschopp E (1961) Zirkulardichroismus und raumstruktur neuere ergebnisse im steroid-gebiet. Angewandte Chemie 73:603–611
6. Moerner WE, Orrit M (1999) Illuminating single molecules in condensed matter. Science 283(5408):1670–1676
7. Tang Y, Cohen AE (2010) Optical chirality and its interaction with matter. Phys Rev Lett 104(16):163901
8. Hendry E, Carpy T, Johnston J, Popland M, Mikhaylovskiy RV, Lapthorn AJ, Kelly SM, Barron LD, Gadegaard N, Kadodwala M (2010) Ultrasensitive detection and characterization of biomolecules using superchiral fields. Nat Nanotechnol 5(11):783–787
9. Zhao Y, Askarpour AN, Sun L, Shi J, Li X, Alù A (2017) Chirality detection of enantiomers using twisted optical metamaterials. Nat Commun 8(1):1–8
10. Nechayev S, Barczyk R, Mick U, Banzer P (2019) Substrate-induced chirality in an individual nanostructure. ACS Photonics 6(8):1876–1881
11. Artega O, Sancho-Parramon K, Nichols S, Maoz BM, Canillas A, Bosch S, Markovich G, Kahr B (2016) Relation between 2d/3d chirality and the appearance of chiroptical effects in real nanostructures. Opt Express 24(3):2242–2252
12. Buckingham AD, Dunn MB (1971) Optical activity of oriented molecules. J Chem Soc A: Inorg, Phys, Theor, pp 1988–1991
13. Schellman J, Jensen HP (1987) Optical spectroscopy of oriented molecules. Chem Rev 87(6):1359–1399
14. Disch RL, Sverdlik DI (1969) Apparent circular dichroism of oriented systems. Anal Chem 41(1):82–86
15. Tang Y, Cook TA, Cohen AE (2009) Limits on fluorescence detected circular dichroism of single helicene molecules. J Phys Chem A 113(22):6213–6216
16. Bohren CF, Huffman DR (2007) Absorption and scattering of light by small particles, appendix a: homogeneous sphere. Wiley, Hoboken, pp 477–482
17. Mulholland GW, Bohren CF, Fuller KA (1994) Light scattering by agglomerates: coupled electric and magnetic dipole method. Langmuir 10(8):2533–2546
18. Banzer P, Woźniak P, Mick U, De Leon I, Boyd RW (2016) Chiral optical response of planar and symmetric nanotrimers enabled by heteromaterial selection. Nat Commun 7(1):1–9
19. Wang L-Y, Smith KW, Dominguez-Medina S, Moody N, Olson JM, Zhang H, Chang W-S, Kotov N, Link S (2015) Circular differential scattering of single chiral self-assembled gold nanorod dimers. ACS Photonics 2(11):1602–1610

20. Smith KW, Zhao H, Zhang H, Sánchez-Iglesias A, Grzelczak M, Wang Y, Chang W-S, Nordlander P, Liz-Marzán LM, Link S (2016) Chiral and achiral nanodumbbell dimers: the effect of geometry on plasmonic properties. ACS Nano 10(6):6180–6188
21. Decker M, Zhao R, Soukoulis CM, Linden S, Wegener M (2010) Twisted split-ring-resonator photonic metamaterial with huge optical activity. Opt Lett 35(10):1593–1595
22. Hentschel M, Ferry VE, Alivisatos AP (2015) Optical rotation reversal in the optical response of chiral plasmonic nanosystems: the role of plasmon hybridization. Acs Photonics 2(9):1253–1259
23. Karst J, Strohfeldt N, Schäferling M, Giessen H, Hentschel M (2018) Single plasmonic oligomer chiral spectroscopy. Adv Opt Mater 6(14):1800087
24. Yuan H, Khatua S, Zijlstra P, Orrit M (2013) Individual gold nanorods report on dynamical heterogeneity in supercooled glycerol. Faraday Discuss 167:515–527
25. Molaei M, Atefi E, Crocker JC (2018) Nanoscale rheology and anisotropic diffusion using single gold nanorod probes. Phys Rev Lett 120(11):118002
26. Xu D, He Y, Yeung ES (2014) Direct observation of the orientation dynamics of single protein-coated nanoparticles at liquid/solid interfaces. Angew Chem Int Ed 53(27):6951–6955
27. Schnoering G, Poulikakos LV, Rosales-Cabara Y, Canaguier-Durand A, Norris DJ, Genet C (2018) Three-dimensional enantiomeric recognition of optically trapped single chiral nanoparticles. Phys Rev Lett 121(2)
28. Trojek J, Chvátal L, Zemánek P (2012) Optical alignment and confinement of an ellipsoidal nanorod in optical tweezers: a theoretical study. J Opt Soc Am A 29(7):1224–1236
29. Sönnichsen C, Reinhard BM, Liphardt J, Alivisatos AP (2005) A molecular ruler based on plasmon coupling of single gold and silver nanoparticles. Nat Biotechnol 23(6):741–745
30. Prikulis J, Svedberg F, Käll M, Enger J, Ramser K, Goksör M, Hanstorp D (2004) Optical spectroscopy of single trapped metal nanoparticles in solution. Nano Lett 4(1):115–118
31. Schreiber R, Luong N, Fan Z, Kuzyk A, Nickels PC, Zhang T, Smith DM, Yurke B, Kuang W, Govorov AO et al (2013) Chiral plasmonic dna nanostructures with switchable circular dichroism. Nat Commun 4(1):1–6
32. Bass M, DeCusatis C, Enoch J, Lakshminarayanan V, Li G, Macdonald C, Mahajan V, Van Stryland E (2009) Handbook of optics, volume II: Design, fabrication and testing, sources and detectors, radiometry and photometry. McGraw-Hill, Inc., New York
33. McQuarrie DA (2000) Statistical mechanics. University Science Books
34. Garcia de la Torre J, Carrasco B (2002) Hydrodynamic properties of rigid macromolecules composed of ellipsoidal and cylindrical subunits. Biopolym: Orig Res Biomol 63(3):163–167
35. Zhang Q, Hernandez T, Smith KW, Jebeli SAH, Dai AX, Warning L, Baiyasi R, McCarthy LA, Guo H, Chen D-H et al (2019) Unraveling the origin of chirality from plasmonic nanoparticle-protein complexes. Science 365(6460):1475–1478
36. Lu X, Wu J, Zhu Q, Zhao J, Wang Q, Zhan L, Ni W (2014) Circular dichroism from single plasmonic nanostructures with extrinsic chirality. Nanoscale 6:14244–14253
37. Ma W, Kuang H, Wang L, Xu L, Chang W-S, Zhang H, Sun M, Zhu Y, Zhao Y, Liu L et al (2013) Chiral plasmonics of self-assembled nanorod dimers. Sci Rep 3:1934
38. Bustamante C, Tinoco I, Maestre MF (1983) Circular differential scattering can be an important part of the circular dichroism of macromolecules. Proc Natl Acad Sci 80(12):3568–3572

Chapter 5
Conclusions and Outlook

"Dem Anwenden muss das Erkennen vorausgehen"
"Insight must precede application"

Max Planck

This thesis investigated the motion, the symmetry and the spectroscopic responses of chiral and achiral nanostructures at low Re. Chirality is important in nature, and as demonstrated throughout this work, it is also an important property of artificial nanostructures. In order to support the increasing role of chiral nanostructures in emerging applications, this thesis has provided a fundamental description of the nanostructures' behaviour in relation to their symmetries. To this end, two universal questions have been examined. First, does a propeller have to be chiral to be propulsive upon rotation-translation coupling and what are the fundamental symmetries to predict the behaviour of simple-shaped propellers? Second, what is the role of symmetry in chiroptical spectroscopy of a single nanostructure and how can its intrinsic chirality undoubtedly be detected?

To clarify the first question, a proposed theory that explains rotation-translation coupling for highly symmetric V-shaped objects at low Re has been experimentally validated. For this, a macroscopic model based on a 3D-printed arc segment as well as GLAD-microstructures were used. Despite their different sizes, both models were shown to be propulsive upon actuation by an external electromagnetic field, depending on the different symmetries the shape together with the dipole moment exhibited. It was demonstrated that the alignment of electromagnetic dipole moments crucially impact the symmetry and thus chirality of the propeller. The theoretically predicted propulsion characteristics were not only confirmed, but based on the experimental findings the existing theory was also extended. Thus it became possible to predict the characteristic non-, bi- or unidirectional propulsion inherent to highly symmetric shapes. Most noteworthy, the first achiral ($\widehat{P}$-even) microswimmer that propels upon

J. Sachs, *Motion, Symmetry & Spectroscopy of Chiral Nanostructures*, Springer Theses,
https://doi.org/10.1007/978-3-030-88689-9_5

rotation-translation coupling was presented. Thereby it was proven that chirality is not a pre-requisite to convert a rotation into a translation at low Re. Confusion due to seemingly achiral swimmers reported in the literature was clarified.

Additionally, experiments with (spherical shaped) swimmers exhibiting very high symmetries and very small sizes ($\ell \lesssim 500$ nm) have been conducted. For this, the active and passive motion of chemically powered Janus *nano*-particle ensembles was observed spectroscopically. Initial experiments with light scattering techniques (DLS, DDM) that rely on the a priori knowledge of a fit-model failed. However, SH-LDV, which is complementary to the others, was identified as a suitable technique for the characterization of the motion of Janus particles, since it is a model-free technique. The enhanced diffusion of the chemically-active sub-micron Janus particles could be shown, including in very dense suspensions. Variation in the shape and size of the particles, as well as delicate changes in the experimental conditions, were shown to impact the measured propulsion velocity distributions of very small self-propelling particles. Preliminary results unraveling the size-dependency of the active motion exhibited by particle ensembles with radii $r \sim 50 - 500$ nm were presented.

The second major focus of the thesis concerned circular differential scattering spectroscopy of single plasmonic nanoparticles. Most prominently, it was demonstrated that symmetries of the shape is of utmost importance for the spectral observation of a single particle, because shape-anisotropic structures in general also exhibit optical anisotropies. Remarkably, even the simplest achiral shapes, like rods, exhibit linear optical anisotropies, which in turn can lead to the misinterpretation of chiroptical spectra. This considerably complicates the spectral analysis of chiroptical responses, because linear polarization artifacts can contribute to the signal, if a stationary single object is measured. To overcome these challenges, a novel experimental approach was developed and implemented, which provides a clean platform for measuring single particle chiroptical spectra and whose interpretation was unambiguous. Thereby the first true chiroptical spectrum of a freely suspended single nanoparticle was measured away from a surface. By taking time-series of spectra and thereby observing a single particle while it statistically samples an isotropic distribution of orientations, the linear polarization effects cancel out. Time-averaging of spectra thus provided a measurement which is only sensitive to the intrinsic chirality of the single particle. This is a major advance for the field of chiral plasmonics, because the time-averaged chiral spectrum is a novel observable that has never been reported elsewhere, and that cannot be obtained from ensemble measurements. Interestingly, the optical detection of chirality only works for motile nanostructures. Further, the ergodicity of chiroptical single-particle spectroscopy was demonstrated for the first time, by showing the similarity between the time-averaged spectra and corresponding ensemble-averaged measurements. It is in general not meaningful to link traditional circular dichroism spectroscopy conducted in a cuvette to measurements on single-particles as long as they are measured stationary. Consequently, the approach presented in this thesis provided the first observation of single-particles, which are in accord with an ensemble of identical particles, but without the need for a large number of particles. Instead, only a single nanoparticle was used.

The special symmetry of chiral nanostructures impacts both motion at low Re and their spectroscopic responses in solution. The interplay between intrinsic properties like symmetry and the mechanical as well as optical properties of chiral nanostructures is anything but trivial and remains a fascinating and important research area. Many more experiments can be thought of for future studies, e.g. actuation of a V-shape that is actuated by means of a precessing field. Similarly, spectroscopy of chemically powered swimmers can help to shed light on the underlying propulsion mechanism and hopefully paves the way for new applications, possibly even biomedical. Experiments concerning the motion of nanostructures were all conducted in Newtonian fluids with a constant viscosity, e.g. water. Biomedical devices need to move in biological tissue which is almost always a non-Newtonian environment, and thus highlights the importance of further experiments in such media. Finally, detection of bio-molecular chirality with the help of plasmonic nanostructures in ultra-low volumes or concentrations down to single-molecule sensitivity is an interesting prospect that has been proposed, but has not been unequivocally demonstrated to date. With the results and the novel approach presented herein, the origin of single-particle plasmonic circular dichroism could be successfully dis-entangled. It is thus feasible that in future also the true chiroptical spectrum exhibited by a single molecule can be observed.

Appendix

A.1 Mueller–Stokes Formalism

The optical response of light propagating through any medium can be modeled mathematically in its most generalized form by utilizing the Mueller calculus. Here, the Mueller–Stokes formalism that has been utilized throughout the thesis is briefly motivated. Especially the matrices being used for calculations are given while a comprehensive description of the formalism can be found in standard textbooks [1]. The linear optical properties of the medium is characterized by this formalism and it is thus used to describe optical components associated with specific optical effects. The full polarization state of the light, including fully, partially and unpolarized contributions can be quantified as a 4-component Stokes vector:

$$\mathbf{S} = \begin{bmatrix} s_0 \\ s_1 \\ s_2 \\ s_3 \end{bmatrix}. \tag{A.1}$$

Any optical element is a 4×4 Mueller matrix $\mathbf{M}$ representing the ability of this element to alter the polarization state of the light. The polarization state after the element is just a matrix multiplication $\mathbf{S}_{\text{out}} = \mathbf{M} \cdot \mathbf{S}_{\text{in}}$, or, by writing it explicitly:

$$\mathbf{S}_{\text{out}} = \begin{bmatrix} s'_0 \\ s'_1 \\ s'_2 \\ s'_3 \end{bmatrix} = \mathbf{M} \cdot \mathbf{S}_{\text{in}} = \begin{bmatrix} m_{00} & m_{01} & m_{02} & m_{03} \\ m_{10} & m_{11} & m_{12} & m_{13} \\ m_{20} & m_{21} & m_{22} & m_{23} \\ m_{30} & m_{31} & m_{32} & m_{33} \end{bmatrix} \begin{bmatrix} s_0 \\ s_1 \\ s_2 \\ s_3 \end{bmatrix} \tag{A.2}$$

The propagation through a series of n elements is then a multiplication of n matrices

$$\mathbf{M}_{\text{Total}} = \mathbf{M_n} \cdot \ldots \cdot \mathbf{M_3} \cdot \mathbf{M_2} \cdot \mathbf{M_1}. \tag{A.3}$$

J. Sachs, *Motion, Symmetry & Spectroscopy of Chiral Nanostructures*, Springer Theses,
https://doi.org/10.1007/978-3-030-88689-9

In order to describe the propagation of light through the dark-field spectroscopy setup presented in Chap. 4 this approach has been used. The individual Mueller matrices used therein were deduced from standard textbooks [1] and are given here explicitly:

$$\mathbf{M_{LP}(0°)} = \frac{1}{2}\begin{bmatrix} 1 & 1 & 0 & 0 \\ 1 & 1 & 0 & 0 \\ 0 & 0 & 0 & 0 \\ 0 & 0 & 0 & 0 \end{bmatrix}, \qquad \mathbf{M_{LP}(90°)} = \frac{1}{2}\begin{bmatrix} 1 & -1 & 0 & 0 \\ -1 & 1 & 0 & 0 \\ 0 & 0 & 0 & 0 \\ 0 & 0 & 0 & 0 \end{bmatrix},$$

$$\mathbf{M_{QWP}(+45°)} = \frac{1}{2}\begin{bmatrix} 1 & 0 & 0 & 0 \\ 0 & 0 & 0 & 1 \\ 0 & 0 & 1 & 0 \\ 0 & -1 & 0 & 0 \end{bmatrix}, \qquad \mathbf{M_{QWP}(-45°)} = \frac{1}{2}\begin{bmatrix} 1 & 0 & 0 & 0 \\ 0 & 0 & 0 & -1 \\ 0 & 0 & 1 & 0 \\ 0 & 1 & 0 & 0 \end{bmatrix}.$$

A.2 Magnetic Coil Setup

For the control of magnetic micro- and nanoparticles with an external magnetic field, a setup consisting of 3 mutually perpendicular Helmholtz coil pairs along x, y and z directions were used. The three coils are controlled independently and thus their fields overlap, which allows to generate a magnetic field with arbitrary direction around the sample. The coil setup was engineered to fit into a standard *Zeiss* microscope stage and in it's center a cover slip (25 mm diameter) could be mounted. The individual coil bodies were machined by the in-house workshop, whereas winding the coils, establishing the electronics as well as writing the control software was done during this thesis. In general the magnetic field of one Helmholtz coil pair in its center point is proportional to

$$B(0) \propto \frac{N \cdot I}{r}, \tag{A.4}$$

where N is the number of windings, I the electrical current and r the coil radius. The magnetic field at the center is thus linear with the driving current. However, a constant current will lead to a constant field only. For rotating the magnetic field vector on an arbitrary trajectory, e.g. on a circle, an alterning current is used. The coils were driven by a computer *via* an analog output device (*Contec* AO-1604LX-USB) that fed amplifiers, which controlled the voltage U. The voltage is dependent on the coil's impedance $|Z|$, and hence:

$$B(0) \simeq I = \frac{U}{|Z|}. \tag{A.5}$$

For an alternating current the impedance is dependent on the driving frequency ω:

$$|Z| \simeq \sqrt{R^2 + (\omega L)^2}. \tag{A.6}$$

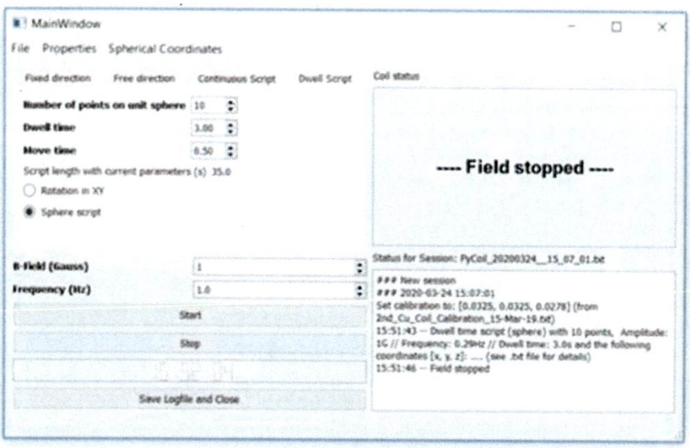

Fig. A.1 Graphical user interface of the custom written python software that controls the magnetic coil setup

Here L is the inductivity and R the resistance of a Helmholtz coil pair in series[1]. Consequently there's a nonlinear relation between $B(0)$ and U for higher B-fields and higher frequencies. However, for low frequencies the varying impedance can be neglect and B is approximately linear with U. Thus, the coils were calibrated by assuming the linear dependency and measuring the magnetic field in practice for several driving currents, i.e. for several magnetic field strengths $B_{\text{theoretical}}$. The calibration coefficient k was then deduced from the slope of the curve $B_{\text{real}} = k \cdot B_{\text{theoretical}}$.

The control software was implemented in a custom written python software comprising a graphical user interface, which is shown in Fig. A.1. It was possible to generate arbitrary field directions and/or various trajectories along which the field vector could be swept. Importantly, this included constant and rotating fields in one plane, but also a (spherical) trajectories on a unit sphere. The latter is described below.

A.2.1 *Spherical Trajectory*

To sample a nanoparticle's alignment over spatially isotropic directions a so called Fibonacci lattice was used. It distributes i points evenly on a unit sphere. The basis is a golden spiral in 2D, which is extruded into the third dimension. The actual implementation in Python is given here.

[1] A LCR bridge measurement revealed $L = 10$ mH and $R = 10\ \Omega$ for the used x and y coils and $L = 20$ mH and $R = 15\ \Omega$ for the two z-spools.

```
def golden_spiral(num_pts=100):
    # ---- Theta, Phi in PHYSICAL/ISO COORDINATES

    #create indices
    indices = np.arange(0, num_pts, dtype=float) + 0.5

    #create theta and phi according to golden spiral method
    theta = np.arccos(1 - 2*indices/num_pts)          #from 0 to Pi
    phi = np.mod((1 + 5**0.5) * indices, 2*np.pi)     #from 0 to 2Pi

    #calculate x,y,z and stack to one vector
    x = np.cos(phi) * np.sin(theta)
    y = np.sin(phi) * np.sin(theta)
    z = np.cos(theta)
    vec = np.vstack((x, y, z)).T.round(15)

    #create a trajectory by connecting "nearest" neighbour points
    v1= vec[1::2]   #rows with odd indices
    v2 = vec[::2]   #rows with even indices
    #reverse stack to avoid jumping
    trajec = np.vstack((v1, v2[::-1]))

    return trajec
```

To avoid random jumping of the magnetic field vector when it moves along the points of a spherical trajectory, adjacent points were connected so that the field vector moved the minimal distance between two points. The resulting theoretical x, y and z coordinates from the `golden_spiral` function are schematically depicted in Fig. A.2a, b as 3D plot as well as their projection into the xy-, xz- and yz-plane. As expected, a smooth trajectory on a unit sphere over adjacent points which are evenly distributed is obtained. After the i evenly distributed points were generated, they were sent to the analog output device, which was connected to the amplifiers driving the the coils. The core of the code which generates and outputs the spherical trajectory with i points is:

```
    ...

    sample_points = self.npoints.value() # number of points on sphere
    script_duration = sample_points/freq
    step_duration = 1/freq
    device.fs=10*freq #physical sampling set to 10 times target sampling rate

    #create a fibonacci lattice
    pt_vec = signal.golden_spiral(int(sample_points))

    #create buffer array with the correct number of samples
    buf_array = np.repeat(pt_vec, device.fs*step_duration, axis=0)
    #multiply with calibration constants and desired field strength
    buf_array = buf_array*self.curCalib*b_field

    #create TTL trigger column
    TTL = np.zeros((buf_array.shape[0], 1))
    #set first value to 5 volts (TTL high)
    TTL[0:int(device.fs*step_duration)] = 5
    #append column with zeros for TTL trigger
    randTTL = np.append(buf_array, TTL, axis=1)

    #rename as buf_array again
    buf_array = randTTL.round(15)

    # set parameters at the analog ouput device and send
```

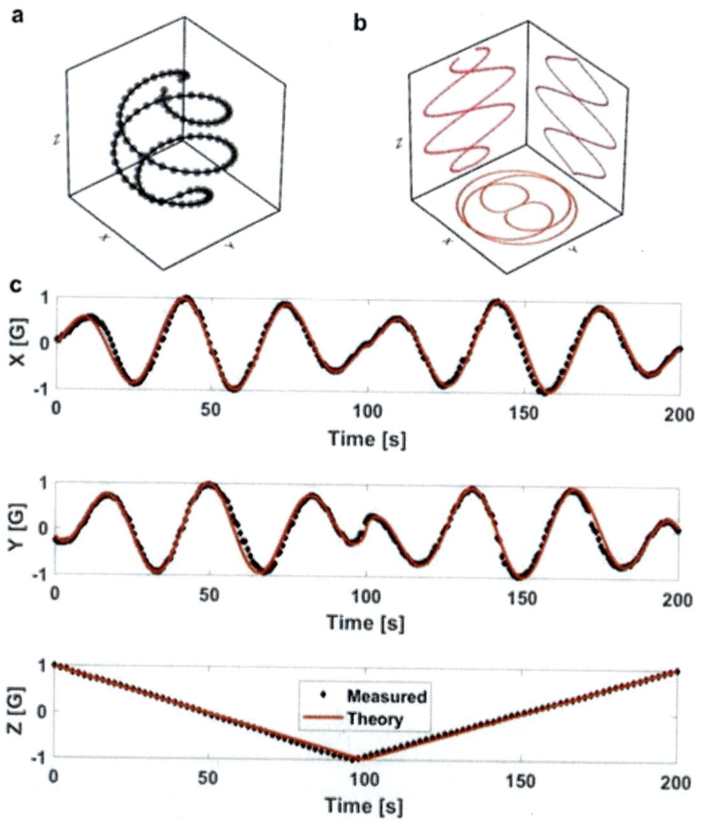

Fig. A.2 "Spherical trajectory" over i evenly distributed points on a sphere. **a** Output of the `golden_spiral` function as 3D plot and **b** the corresponding projection into the xy-, xz- and yz-planes respectively. **c** Theoretical and measured x-, y- and z-coordinate of the magnetic field vector moving on a continuous spherical trajectory with $i = 200$ and a sampling rate of 1 point per second

```
    device.n_channels = 4           #set 4 output channels
    device.buffer = buf_array       #write lattice point to buffer
    device.start()                  #start the analog output

    ...
```

Together with the x, y and z coordinates a TTL trigger pulse was appended to another analog output channel, which was used to synchronize the timing of the script with the spectrometer, e.g. it's starting time. The code excerpt presented here creates a continuously varying field vector. For particles that were immersed in a highly viscous medium, the exerted magnetic torque was potentially low due to the small magnetic dipole moment. Hence, the nanoparticle will probably not align instantaneously with the external field vector. It was thus desirable to first move the external field and wait for the alignment to happen, before a spectrum was acquired. Such timing patterns were also considered in the software. This is seen in Fig. A.1, where the parameter `Move_time` is the waiting time after the field has been moved and `Dwell_time` specifies the exposure time of a spectrum during which the field vector was held constant. Consequently, the field vector moved to a specific orientation in $\{x, y, z\}$ and a TTL pulse, which opened the shutter of the CCD detector, was sent after waiting for a time specified by the `Move_time`. Measurements of magneto-plasmonic nanostructures' CDSI spectra that were swept along a spherical trajectory rely on this procedure (Sect. 4.5.2).

To verify the proper functioning of the software, the magnetic field was measured with a Gauss-Meter in the center of the 3-axis Helmholtz coil. The result is plotted against the theoretical vector coordinates for all three spatial directions (see Fig. A.2c). The agreement is within the accuracy of the measurement device and validates that the magnetic field vector was moving along a spherical trajectory that samples spatially isotropic orientations.

A.3 Single Object Tracking—Macroscopic Body (Python)

The experiments of macroscopic 3D-printed arc-segments were recorded with a commercial camera (*Canon* EOS 600D). To deduce the propulsion velocity the arcs' position was tracked in every frame of the video. This was done by batch processing the videos, which was convenient because many videos at different rotation frequencies were recorded. The code for this was written in python and relies on the openCV library which provides various real-time computer vision functions. Briefly, a colour filter is applied to a single frame to deduce a two-colour image (black & white). For this, the arc was physically painted with red colour and consequently the algorithm produces the arcs' body (white) in front of a black background. A comparison between a raw frame and the filtered image is shown in Fig. A.3. Then, in the binary image the contour of the white patch is detected, it's center point calculated and the

Fig. A.3 Object tracking of an object that has a different color than the background. A color filter was applied to the raw image (left), blackening any pixel which is not red and creating a binary image (right). The latter was then utilized for finding the objects' position in one frame

data saved to a list, before the next frame is processed. In the end, the list that contains the positional and timing data is saved as a text file, which was later opened by other software programs to plot the arcs' coordinates over time and retrieve the velocity. For convenience the code comprises also some command line output to inform the user which video is being processed and it could be aborted by pressing the button "q".

Videos acquired during different experimental sessions had different environmental variables. The background colour changed or the arc was colored not only red but also one end green (see Fig. 3.6). Thus, the executed code was usually adjusted slightly to achieve the best results, but the core remained the same and is presented here:

```
import cv2
import numpy as np
import os
import datetime

# Determine red color which is tracked
# CorelDraw HSV uses H = 0-360,  S = 0-100   and V = 0-100.
# But OpenCV uses    H = 0 - 180, S = 0 - 255 and V = 0 - 255
RED_MIN = np.array([0, 10, 10],np.uint8)
RED_MAX = np.array([20, 250, 250],np.uint8)

# set the artifact threshold in pixel size
artefacts = 100

stopbutton = 0
cv2.namedWindow('Tracking in process')

#set working directory
dir = '\\ zwe\pfgroup\+ PEOPLE\Johannes\Arc result'
os.chdir(dir)

#get all files in folder with ending .MOV
files = []
for file in os.listdir(os.getcwd()):
    if file.endswith(".MOV"):
        files.append(file)

#start tracking for actual video
count_vids = 1 #initialize counter for videos

for actual_vid in files:
    start_time = datetime.datetime.now()
    #check if q was pressed, if so then exit program
    if stopbutton == 1:
        break

    #get name of video and print message to user
    basename = actual_vid.rsplit( ".", 1 )[0]
    print( 'Video {} of {}'.format(count_vids, len(files)) +
           ' is being processed ({})'.format(actual_vid) + '\n'
           'Please wait or abort by pressing q ...' )

    #open video and get some information
    cap = cv2.VideoCapture(actual_vid)
    total_frames = cap.get(cv2.cv.CV_CAP_PROP_FRAME_COUNT)
    fps = cap.get(cv2.cv.CV_CAP_PROP_FPS)


    without_object = 0  # counter for frames w/o object
    # initialize list for every quantity which is extracted to file
```

```
    time_s = []
    frame_nr = []
    coord_x = []
    coord_y = []

    while True:
        # Read a frame of the video
        grabbed, frame = cap.read()
        key = cv2.waitKey(1) & 0xFF

        # if no frame is grabbed, then end of the video reached
        if not grabbed:
            print "Processing␣" + actual_vid + "␣ended"
            count_vids += 1
            break

        # get actaul frame and playtime
        f_nr = cap.get(cv2.cv.CV_CAP_PROP_POS_FRAMES)
        seconds = cap.get(cv2.cv.CV_CAP_PROP_POS_MSEC)/1000

        # process frame
        hsv_img = cv2.cvtColor(frame,cv2.COLOR_BGR2HSV)
        frame_threshed = cv2.inRange(hsv_img, RED_MIN, RED_MAX)
        # dilate the thresholded image and find contours
        thresh = cv2.dilate(frame_threshed, None, iterations=2)
        contours, hierarchy = cv2.findContours(thresh,cv2.RETR_EXTERNAL,
                                    cv2.CHAIN_APPROX_SIMPLE)


        # find largest contour
        if (len(contours) > 0):
            #create array for contours
            areaArray = []
            for i, c in enumerate(contours):
                area = cv2.contourArea(c)
                areaArray.append(area)

            #first sort the array by area
            sorteddata = sorted(zip(areaArray, contours), key=lambda x: x[0],
                                reverse=True)
            #find the nth largest contour [n-1][1], in this case the 1st
            largestcontour = sorteddata[0][1]

            # prevent division by zero
            M = cv2.moments(largestcontour)
            if (M["m00"] == 0):
                M["m00"]=1
            #calculate centerpoint of contour
            cX = int((M["m10"] / M["m00"]))
            cY = int((M["m01"] / M["m00"]))

            #filter artefacts and store the centerpoint and related data
            if cv2.contourArea(largestcontour) > artefacts:
                coord_x.append(cX)
                coord_y.append(cY)
                frame_nr.append(f_nr)
                time_s.append(seconds)

        else: # if no contour was found = no red object in frame
            without_object += 1
        if key == ord('q'): # quit process by pressing q
            stopbutton = 1
            break

    cap.release() #end processing of actual video
    head = ('....␣HEADER␣DATA␣....') # specify header for output_file
    # save data to file
```

```
    np.savetxt(basename + '.txt', np.column_stack(
                    (time_s, frame_nr, coord_x, coord_y)),
                    delimiter=";", fmt='%s', header = head)
cv2.destroyAllWindows() #close openCV
```

Initially the python libraries are loaded and some parameters are defined before the videos in the specified folder are read. The code then opens a video, retrieves some metadata like the frame rate and prints command line outputs. The colour filter is applied via simple thresholding. After that the binary image is dilated to fill in holes between smaller white patches and to get rid of smaller artifacts arising in other regions of the image. For instance the binary image in Fig. A.3 contains small white dots located over the arcs contour, whereas the contours itself consists of many smaller tiles. A dilatation will close the holes between them and results in vanishing artefacts. Then the `cv2.findContours` function detects the contours in the image and it was assumed that the biggest contours corresponds to the arc. Therefore the contours are sorted from large to small and the center point coordinates were calculated for the largest. In some frames the magnetic disk blocks the camera view and no contour was detected, hence those frame were skipped. However, sometimes the color filtering produced a binary image which contained small white dots that were occurring randomly. Consequently the algorithm detected those artefacts, and to account for this, the algorithm saves only contours that exceed a certain size (specified by the parameter `artefacts`). Nevertheless, in some frames this doesn't help either and the code detects and saves an erroneous position, which doesn't correspond to the arc's position. This is the reason for huge jumps observed when the position was plotted over time, e.g. for the data shown in Fig. 3.4. Such outliers will cancel out when the object is watched for sufficiently long times and a linear regression is fitted.

A.4 Multi-object Tracking-Microscope Videos (Python)

Micro- and nanostructures fabricated by GLAD were observed under a microscope and videos were acquired by using either an electron multiplying CCD (*Andor* iXon) or a scientific complementary metal–oxide–semiconductor camera (*Andor* Zyla 4.2 & 5.5). To extract the frame-wise location of multiple objects again python was used, similar to the macroscopic videos. However, a simple color thresholding for multi-object recognition usually fails, unless the objects exhibit the identical color. Additionally, the cameras acquire only gray-scale videos, which exacerbate problems with color filtering. However, deducing dynamic objects in a video is relatively simple if the background is subtracted. In doing so, a binary image containing only moving objects is obtained. Thus, initially the background of an acquired video was deduced. For this, every individual pixel is averaged over all frames:

```
cap = cv2.VideoCapture('movie.avi')
while True:
    grabbed, frame = cap.read()
    frame_nr = cap.get(cv2.CAP_PROP_POS_FRAMES)
    if not grabbed:
        print "Processing " + fname + " ended"
        break

    img_gray = cv2.cvtColor(frame, cv2.COLOR_BGR2GRAY).astype('float64')
    cv2.accumulate(img_gray,avg)
    backg = avg/frame_nr
# save when done accumulating
im = Image.fromarray(backg)
im.save('backg.jpg')
```

The idea behind is that a moving object is present at a specific pixel for a short time only. Consequently the average pixel value is practically only determined by a (static) background, whereas the impact of a particle that moves over this pixel is vanishing for sufficiently long observation times. The averaged background is then subtracted from an individual frame to obtain again an image in which the contours can be detected. In Fig. A.4a the raw signal of a random frame in a microscope video is shown. If the acquired background (Fig. A.4b) gets subtracted only the objects which change their position over the course of the video remain as a bright spot as shown in Fig. A.4c. The corresponding code is:

```
cap = cv2.VideoCapture('movie.avi')
backg = cv2.imread('backg.jpg',0)
kernel = np.ones((4,4))

while True:
    grabbed, frame = cap.read()
    key = cv2.waitKey(1) & 0xFF
    if not grabbed:
        print "Processing " + fname + " ended"
        break

    img_gray = cv2.cvtColor(frame, cv2.COLOR_BGR2GRAY).astype('float64')
    bgfree = cv2.subtract(img_gray,backg.astype('float64'))
```

The background-free image can then be further processed by applying a thresholding (white color) to obtain a binary image, which is then easy to evaluate, e.g. with the above mentioned `cv2.findContours` command of the openCV library. An alternative is the trackpy library providing the `trackpy.locate` function. The latter is especially useful because it relies on the pandas data structuring library that is advantageous for handling large data sets. If a single object is observed it is clear that the bright spot and/or the largest contour in the binary images corresponds to the target object. Conversely, during multi-object tracking it's not readily obvious if a detected feature belongs to one and the same particle over subsequent frames. Thus, after recognizing feature locations in a single frame, they have to bey linked to a trajectory over the course of the video. In this thesis this was done with the `trackpy.link_df` command. The trajectories were then evaluated to derive the particles's velocities.

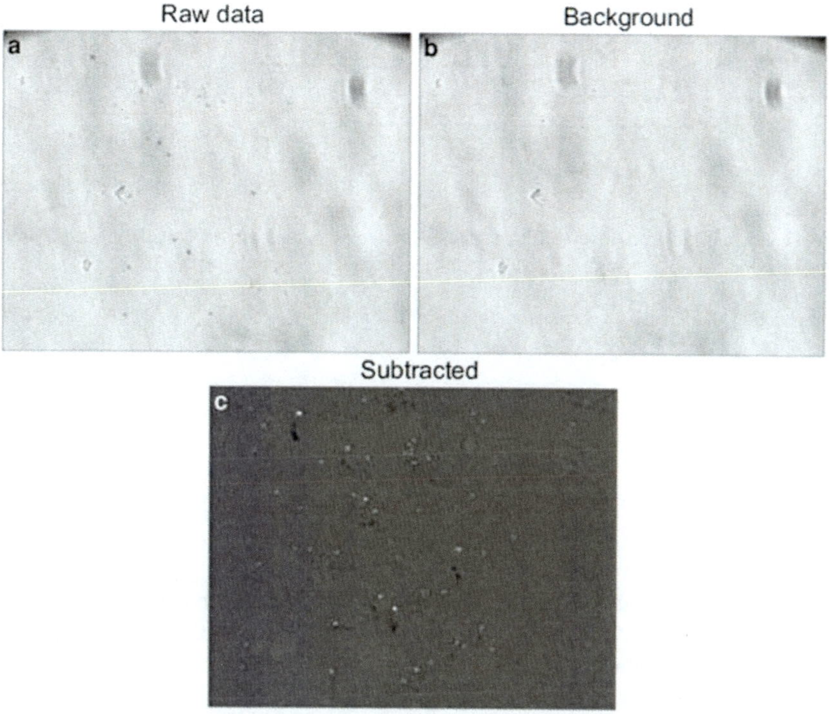

Fig. A.4 Background subtraction of a microscope video containing moving particles. **a** Raw image. **b** Averaged background over 100 frames. **c** Background-subtracted image of the raw data shown in **a**

A.5 Differential Dynamic Microscopy (Matlab)

The DDM technique is complementary to DLS because both are employed to measure characteristic diffusion of a particle ensemble and the physics behind is the same. An intensity is recorded over time $I(t)$ and the characteristic fluctuations get analyzed. However, DDM uses standard bright-field microscopy with a camera rather than a detector at one scattering angle. Also, no hardware correlator is used. Instead, in DDM the individual images of a captured video get correlated, and therefore the intensity fluctuation exhibited by individual pixels is evaluated. Compared to DLS much smaller **q**-vectors are accessible and interestingly multiple **q**-vectors are recorded in one video. The technique was introduced in 2008 [3] and was used for studying the dynamics of many different systems, including Janus particles [4, 5].

Figure A.5 schematically visualizes the DDM-algorithm that was implemented during this thesis, whereas the exact description of the method is beyond the scope here. Two reports about the application of the technique with bright- and dark-field microscopy explain the details [2, 6]. In fact, the code presented here was inspired by those two references that gave a comprehensive and easy to follow protocol of DDM.

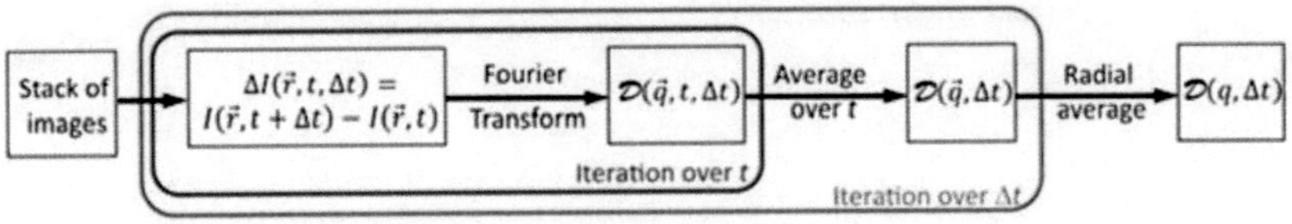

Fig. A.5 Schematic sketch showing the working principle of the DDM algorithm which runs over two loops: One over different t averaging over several image pairs whose separation is Δt, and second one that runs over different spacing Δt of the images. By calculating the Fourier transform and radially averaging the camera pixels, the image structure function $\mathscr{D}(\mathbf{q}, t, \Delta t)$, which is proportional to the intermediate scattering function is obtained. Reproduced from [2], with the permission of the American Association of Physics Teachers

Briefly, a time series of images yields an array of pixels with spatial coordinate $\mathbf{r}$, thus each pixel has an intensity $I(\mathbf{r}, t)$. The difference between two images separated by a lag time $\Delta t = t_2 - t_1$ is:

$$\Delta I(\mathbf{r}, t, \Delta t) = I(\mathbf{r}, t + \Delta t) - I(\mathbf{r}, t). \tag{A.7}$$

Image subtraction cancel out static contributions, whereas the dynamics exhibited by an object which varied its position between the two images is still present. A Fourier transform is then executed on the difference image $\text{FFT}\{\Delta I(\mathbf{r}, t, \Delta t)\} = |\widehat{\Delta I}|$, which yields the image structure function $\mathscr{D}(\mathbf{q}, t, \Delta t) = \left\langle |\widehat{\Delta I}|^2 \right\rangle_t$. The t denotes an average over a large number of image pairs exhibiting a time distance of Δt, which is necessary to achieve a reasonable averaging over the dynamics happening at this lag-time. In the end a radial averaging over the camera pixel array is done because pixels corresponding to the same scattering vector $\mathbf{q} = \sqrt{q_x{}^2 + q_y{}^2}$ provide the same information. It follows that the scattering vector reduces to a scalar. Afterwards, the same calculations and averaging is repeated for another lag time Δt_2 and so on. In the end the DDM-matrix or image structure function $\mathscr{D}(q, \Delta t)$ is obtained:

$$\mathscr{D}(q, \Delta t) = A(q)[1 - f(q, \Delta t)] + B(q), \tag{A.8}$$

which can be identified with a signal $A(q)$, and the noise $B(q)$ the camera was recording. It depends on the scattering vector and represents the dynamics happening within the observed video, i.e. it is proportional to the intermediate scattering function $f(q, \Delta t)$. This function is the same as the intensity autocorrelation function measured by DLS $g_2(\tau) = f(q, \Delta t)$, which underpins that both methods are complementary. From the computational point of view, the DDM algorithm runs over two loops. One over t that averages several image pairs separated by Δt, while the second loop runs over Δt computing the image structure function for a different lag-time.

The advantage of DDM is its simplicity. Beside a microscope which can be low-tech, only a computer with sufficient calculation power is needed for processing the algorithm. During this thesis DDM was implemented in Matlabb for the evaluation of microscope videos acquired with an scientific complementary metal-oxide-semiconductor camera (*Andor* Zyla). To sustain high frame rates (>1000 fps) but also to save time, the videos were conveniently saved in a proprietary file format (*.sifx*). Instead of doing a time-consuming conversation of the videos into an image-series, it was advantageous to process them directly in the given format. In order to do so, several subroutines have been written for handling the files which rely on the Andor Software Development Kit. Most notably is the function `DDM_grabframe` which deduces an image of a video file and returns it as an array of pixels:

```
function [ img ] = DDM_grabframe( fnumber, signal, fsize, width, height)
[rc,tmp]=atsif_getframe(signal,fnumber,fsize);
if rc~=22002 %if not 22002 somethign went wrong
    pause(0.5) %retry after a short wait period
    [rc,tmp]=atsif_getframe(signal,fnumber,fsize);
    if rc~=22002 %if still no data can be deduced, throw an error
        error('Error.␣ \nInput␣must␣be␣a␣char.␣\nReturn␣code:␣%s␣\n#␣of␣frame␣that␣was␣tried␣to␣open:␣%i␣\nTime:␣%s', num2str(rc), num2str(fnumber) ,datestr(now));
    end
end
    img = reshape(tmp,width,height);
end
```

This function was then used in the main program which opened a *.sifx* video file and consecutively grabbed the frames needed for the DDM algorithm. Averaging over every frame of a video would have led to enormous calculation times. Therefore some shortcuts which limit the calculations time have been introduced. First, not every lag-time Δt was computed. Instead usually a logarithmic scaling between $\Delta t_1, \Delta t_2, \ldots$ was used to limit the number of time-steps. If necessary the spacing was changed to a linear scale. Moreover, the inner loop which runs over t was limited to a fixed number of images that contribute (parameter `avgframes`). Despite those limitations which reduced the calculation time significantly, the computation time for processing a single video was still lengthy. It was thus convenient to do a batch processing of several videos over night (not shown here). Therefore the script also deduces some metadata like the frame rate, pixel array size, etc. for an automatic processing. After the DDM calculation was done, the metadata was used to further calculate the real units of the scattering vectors ($\left[\frac{1}{\mathrm{m}}\right]$) and timing information according to the used frame rate and pixel calibration of the microscope. Importantly, the algorithm for calculating the Fourier transform only worked for a square array of pixels and hence rectangular arrays were cropped accordingly. The main program is given here:

```
function [DDM, realq, realdt, parameters, infmsg] = DDM_script(filein, pxsz, limit_t, calculationtime)
%open file
    [ path, parameters] = DDM_openfile(filein);
    %parameters = {signal, fsize, width, height, cycle_time, no_frames}

%define timesteps to be computed
    if calculationtime==0 %linear spacing
        timesteps=round(linspace(1,15,9)); %time steps (deltaT) to be computed (outer loop)
    else %log spaced timesteps according to framerate
        timesteps=unique(round(exp(0:0.08:log(parameters{6}))));
    end
```

```
    if any(timesteps>parameters{6})
        ME = MException('MyComponent:noSuchVariable', 'timesteps vector wrong, exceed max number of frames');
        throw(ME);
    end

% deduce metadata
        metad = header(0); %retrieve metadata
        msg = sprintf('%s\nFile Info for: %s\n%i frames with %ix%i\n', datestr(now), filein, parameters{6}, parameters{3}, parameters{4});
        msg2 = sprintf('%s %s\n', metad{6}, metad{4}, metad{7:8}, metad{9:10}, metad{19:20});
        mb = msgbox(char(msg,msg2), 'DDM calculation started', 'help');
% if the image is not a square define the region to crop
        smallerdim=min(parameters{3}, parameters{4});  %length of smaller dimension
        if parameters{3} =parameters{4}
            imgsize = 1:smallerdim;
        else
            imgsize = 1:parameters{3};
        end

% define some variables
        DDM = []; %initialize ddm output vector
        wb = waitbar(0,sprintf('Start processing ... '), 'Name','Progress'); %show a waitbar
        avgframes=parameters{6}-timesteps;  %use maximum iterations for inner loop (max frames-timestep)
% do calculation
%------- OUTER LOOP
    for delta=1:size(timesteps,2)       %iteration over deltaT
        dt=timesteps(delta)*parameters{5};  %dt is the real time step in ms, depending on fps(parameters{5}=cycle_time)
        waitbar((delta/size(timesteps,2)),wb,sprintf('Processing loop %i of %i (\\Deltat= %.2d s)', delta, size(timesteps,2), dt));
        sumPS =0; %initialize summedPowerSpectrum variable
        divisor=avgframes(delta); %to normalize the avgPS later

        if avgframes(delta)>limit_t %define how many frames should be processed for the actual delta
            fr2process=limit_t;                     %limit inner iteration
        else
            fr2process=avgframes(delta);     %or use maximum timesteps
        end
%------- INNER LOOP
        for k=0:fr2process
            %get frames and calculate difference image
            Di_r = DDM_grabframe(k+delta,parameters{1:4})-DDM_grabframe(k,parameters{1:4}) ;
            %calculate power spectrum for difference image
            Fd_q = abs(fftshift(fft2(Di_r(imgsize, imgsize)))); % fft not normalized by size
            %sum up power spectra for different t but same \delta t
            sumPS = sumPS+Fd_q.^2;
        end

        avgPS = sumPS./fr2process; %normalize Power Spectrum
        [avgEXCNT, ~] = DDM_excnt(avgPS,1); %exclude artifical values near center (with q~0)
        %calculate radial average over q
        [PS, q] = DDM_radialavg(avgPS, floor(parameters{3}/2));
        %save Power spectrum and step
```

```
        DDM(:, delta)=PS;
% Each column in DDM represents D(q,dt)as a function of q at a
    fixed time step dt;
% each row represents D(q,dt) as a function of time step dt at a
    fixed q.
    end
    delete(wb);

%calculate lag time and q in real units
    realdt = timesteps*parameters{5}; %dt are the real time steps
        in s depending on fps! (parameters{5}=cycle_time)
    sz = min(parameters{3}, parameters{4}); %get side length of
        image in px
    realq = (0:(sz-1)/2)*2*pi/(sz*pxsz);  %in 1/um. because of
        radial avg only q along one half of the image

    atsif_closefile(); %close
    msg3=sprintf('%s\n%s\nDDM␣calculation␣completed!', datestr(now
        ), path);
    mb = msgbox(msg3, 'DDM␣calculation␣successful', 'Success');
    infmsg= char(msg,msg2,msg3);
end
```

After $\mathscr{D}(q, \Delta t)$ was obtained it had to be fitted, similar to the autocorrelation-functions in DLS. To test the Matlab code, videos with passive SiO_2 beads that were randomly moving due to Brownian motion were evaluated. The fit model for Brownian motion is known ($\Gamma = -D_\tau q^2$) and the advantage of DDM is that the **q**-dependency is automatically accessed. Therefor, a simple linear regression of Γ vs. q^2 was done subsequently and is shown in Fig. A.6. The silica beads investigated here had the same sizes that have been used for the Janus particle experiments in Fig. 3.14. The theoretical diffusion constants D calculated with Eqs. (2.15) and (2.16) (for $\eta = 1$ cP and $T = 298$ K) are:

$$
\begin{aligned}
D &= 3.00 \cdot \mu\mathrm{m}^2\mathrm{s}^{-1} \quad \text{for} \quad r = 72\,\mathrm{nm},\\
D &= 1.66 \cdot \mu\mathrm{m}^2\mathrm{s}^{-1} \quad \text{for} \quad r = 130\,\mathrm{nm},\\
D &= 1.22 \cdot \mu\mathrm{m}^2\mathrm{s}^{-1} \quad \text{for} \quad r = 177\,\mathrm{nm},\\
D &= 0.83 \cdot \mu\mathrm{m}^2\mathrm{s}^{-1} \quad \text{for} \quad r = 259\,\mathrm{nm},\\
D &= 0.38 \cdot \mu\mathrm{m}^2\mathrm{s}^{-1} \quad \text{for} \quad r = 575\,\mathrm{nm}.
\end{aligned}
\tag{A.9}
$$

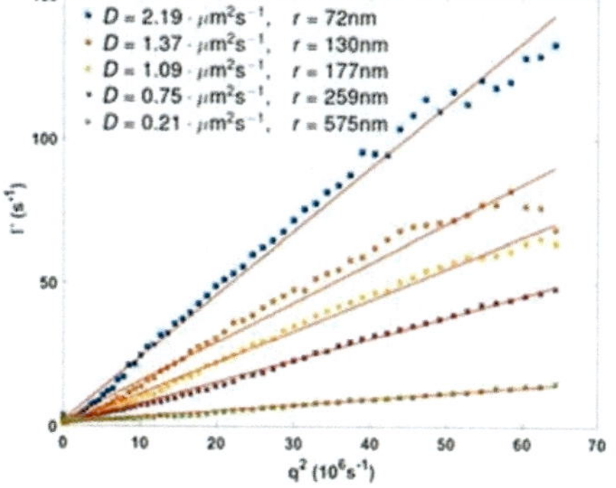

Fig. A.6 DDM of passive Brownian silica particles. The diffusion constant for differently sized spheres is obtained by fitting the decorrelation time Γ against the scattering vector q^2 after videos have been evaluated with the DDM algorithm

Although the experiments and especially the fitting was done without special care, the theoretical and experimentally extracted values for the diffusion coefficients differ only slightly and thus confirmed a proper functioning of the algorithm. However, a proper fitting of $\mathscr{D}(q, \Delta t)$ failed for experiments with Janus particles because the underlying diffusion model is lacking. Also, a comparison of the raw $\mathbf{q}$-dependency for passive and active Janus particles did not yield reliable results.

References

1. Bass M, DeCusatis C, Enoch J, Lakshminarayanan V, Li G, Macdonald C, Mahajan V, Van Stryland E (2009) Handbook of optics, volume II: design, fabrication and testing, sources and detectors, radiometry and photometry. McGraw-Hill, Inc., New York
2. Germain D, Leocmach M, Gibaud T (2016) Differential dynamic microscopy to characterize brownian motion and bacteria motility. Am J Phys 84(3):202–210
3. Cerbino R, Trappe V (2008) Differential dynamic microscopy: probing wave vector dependent dynamics with a microscope. Phys Rev Lett 100(18):188102
4. Cerbino R, Cicuta P (2017) Perspective: differential dynamic microscopy extracts multi-scale activity in complex fluids and biological systems. J Chem Phys 147(11):110901
5. Wittmeier A, Leeth Holterhoff A, Johnson J, Gibbs JG (2015) Rotational analysis of spherical, optically anisotropic janus particles by dynamic microscopy. Langmuir 31(38):10402–10410
6. Bayles AV, Squires TM, Helgeson ME (2016) Dark-field differential dynamic microscopy. Soft Matter 12(8):2440–2452